The Rhetorical Construction of Vegetarianism

This book explores themes in the rhetoric of vegetarian discourse. A vegan practice may help mitigate crises such as climate change, global health challenges, and sharpening socioeconomic disparities, by ensuring both fairness in the treatment of animals and food justice for marginalized populations. How the message is spread is crucial for these aims.

Vegan practices thus uncover tensions between individual dietary choices and social justice activism, between ego and eco, between human and animal, between capitalism and environmentalism, and within the larger universe of theoretical and practical ethics. The chapters apply rhetorical methodologies to understand vegan/vegetarian discourse, emphasizing, for example, vegan/vegetarian rhetoric through the lens of polyphony, the role of intersectional rhetoric in becoming vegan, as well as ecofeminist, semiotic, and discourse theory approaches to veganism. The book aims to show that a rhetorical understanding of vegetarian and vegan discourse is crucial for the goals of movements promoting veganism.

The book is intended for a wide interdisciplinary audience of scholars, researchers, and individuals interested in veganism, food and media studies, rhetorical studies, human-animal studies, cultural studies and related disciplines. It urges readers to examine vegan discourses seriously, not just as a matter of personal choice or taste but as one vital for intersectional justice and our planetary survival.

Cristina Hanganu-Bresch is Associate Professor of English at St. Joseph's University. Her research so far has focused on the rhetoric of health and medicine, with a focus on psychiatry, and scientific writing; she also has a strong interest in animal studies and vegetarianism. She has co-authored and edited several books, among which *Diagnosing Madness* (with Carol Berkenkotter, 2018), and *Veg(etari)an Arguments In Culture, History, and Practice: The V Word* (with Kristin Kondrlik, 2021).

Critical Food Studies
Series editors: Michael K. Goodman, *University of Reading, UK,* and **Colin Sage,** *Independent Scholar*

The study of food has seldom been more pressing or prescient. From the intensifying globalisation of food, a world-wide food crisis and the continuing inequalities of its production and consumption, to food's exploding media presence, and its growing re-connections to places and people through 'alternative food movements', this series promotes critical explorations of contemporary food cultures and politics. Building on previous but disparate scholarship, its overall aims are to develop innovative and theoretical lenses and empirical material in order to contribute to – but also begin to more fully delineate – the confines and confluences of an agenda of critical food research and writing.

Of particular concern are original theoretical and empirical treatments of the materialisations of food politics, meanings and representations, the shifting political economies and ecologies of food production and consumption and the growing transgressions between alternative and corporatist food networks.

Metaphor, Sustainability, Transformation
Transdisciplinary Perspectives
Edited by Ian Hughes, Edmond Byrne, Gerard Mullally and Colin Sage

Food and Cooking on Early Television in Europe
Impact on Postwar Foodways
Edited by Ana Tominc

Hunger and Postcolonial Writing
Muzna Rahman

Food Sovereignty and Urban Agriculture
Concepts, Politics, and Practice in South Africa
Anne Siebert

The Rhetorical Construction of Vegetarianism
Edited by Cristina Hanganu-Bresch

For more information about this series, please visit: www.routledge.com/Critical-Food-Studies/book-series/CFS

The Rhetorical Construction of Vegetarianism

Edited by Cristina Hanganu-Bresch

Routledge
Taylor & Francis Group
LONDON AND NEW YORK

First published 2023
by Routledge
4 Park Square, Milton Park, Abingdon, Oxon OX14 4RN

and by Routledge
605 Third Avenue, New York, NY 10158

Routledge is an imprint of the Taylor & Francis Group, an informa business

British Library Cataloguing-in-Publication Data
A catalogue record for this book is available from the British Library

ISBN: 9780367482794 (hbk)
ISBN: 9781032448657 (pbk)
ISBN: 9781003039013 (ebk)

DOI: 10.4324/9781003039013

Typeset in Times NR MT Pro
by KnowledgeWorks Global Ltd.

Contents

Contributors

Lora Arduser, PhD, Associate Professor and Associate Professor & Director of Professional Writing, Department of English, University of Cincinnati.

Francesco Buscemi, PhD, Contract Professor, Università Cattolica del Sacro Cuore, Università dell'Insubria.

Abby Dubisar, PhD, Associate Professor, Department of English, Iowa State University.

Silke Feltz, PhD, Assistant Teaching Professor, University of Oklahoma.

Lindsay Garcia, M.F.A, Ph.D., Assistant Dean of the College for Junior/Senior Studies, Adjunct Assistant Professor of American Studies, Brown University.

Sibylle Gruber, PhD, Professor and Area Chair for Rhetoric, Writing, and Digital Media Studies, Northern Arizona University.

D. R. Hammontree, PhD, Assistant Professor, Department of Writing and Rhetoric, Oakland University.

Cristina Hanganu-Bresch, PhD, Associate Professor, Department of English, Writing and Journalism, St. Joseph's University.

Kristin Kondrlik, PhD Associate Professor of English, Co-Director of the Professional and Technical Writing Minor, Department of English, West Chester University of Pennsylvania.

Erin Trauth, PhD, Assistant Professor of English, High Point University.

1 The Discursive Construction of Veg(etari)anism

A Polyphonic Rhetoric

Cristina Hanganu-Bresch

We live in a world of natural wonders and amazing diversity in terms of species—the only planet we know, so far, capable of sustaining such a flourishing variety of lifeforms. Entire areas of our sciences, humanities, media, and the arts are devoted to examining, understanding, and appreciating our natural world. There is, as the saying goes, no Planet B that could support our life as we know it.

We have also set this world on fire[1] by polluting it and pushing consumption and exploitation of natural resources (including living beings, human and nonhuman alike) to a point where all that we appreciate about life on earth is now imminently threatened by a variety of catastrophic disasters that are bound to change all biota, and all human and animal futures. An important component of climate change and environmental pollution has been conclusively linked to industrial animal farming, driven by an insatiable, state-sanctioned demand for meat, eggs, and cheese; deforestation, greenhouse gas emissions, and ocean and waterway pollution are just some of the consequences of factory farms and industrial fishing. At a minimum, 15% of anthropogenic greenhouse emissions causing global warming are due to industrial animal farming; consequently, we cannot address global warming without addressing industrial agriculture (FAO, 2013). However, vast segments of our economic and political lives are devoted to ensuring this sort of exploitation continues.

We exist, today, in a state bookended by these two opposing sentiments: marvel and reverence for all life, and despair triggered by the continuous destruction of the same.

This is when discussions about veganism/vegetarianism take on a new life and a new impetus, and in a way mirror this chasm. Vegans will often point out the cognitive dissonance of omnivores who will profess respect and even love for some nonhuman animals just as they partake in the consumption of different types of nonhuman animals at the communal table, animals whose lives were short, painful, and ended in agony. Love and reverence for nature and life in general seem to end at the dinner table for most people. This cognitive dissonance parallels the one many face when confronted with the reality of climate change: recognizing that our actions, collectively, worsen

DOI: 10.4324/9781003039013-1

the climate crisis while feeling helpless to stop, or even bound to continue unsustainable practices. Veg(etari)anism is a millennia-old practice, certainly much older than the industrial revolution, the point in time when human activities started accelerating our climate change crisis. However, in the wake of our current environmental emergency, veganism takes more urgent and even deeper existential meaning than it has ever had. In essence, veg(etari)anism asks us who we want to be now—how do we align our moral values with our praxis—and how do we want our future to be, as the current trajectory points to cascading and irreversible ecological upheaval on a planetary scale.

The discursive rhetoric of veg(etari)anism is nowadays forged in the interstitial chasm between these two poles: on the one hand, clinging to and continually legitimizing a "traditional" way of doing things, and on the other recognizing the inevitability of climate change and related current and near-future catastrophes. Arguments for adopting vegetarianism/veganism have a rich history—but given our ecological crises, they acquire new urgency and need to be redefined accordingly. Today's vegan movements engage in a polyphonic or chameleonic rhetoric, which aims to recognize and embrace all the perceived motivations for meat eating and meat abstention, and to respond to those motivations on multiple levels and with a variety of arguments that are sometimes incommensurable, as they reflect the vast variety of audiences they are trying to reach.

I used the plural "movements" because there is no unified bloc, but multiple factions loosely united by the "vegan" appellation—and sometimes their interests may compete: in the United States, they range from PETA and similar animal liberation groups to seven-day Adventists to fringe nutrition-focused groups and everything in between. Not all vegans would endorse all of PETA's public arguments for going vegan, which in the recent past have often made use of overtly sexist imagery or of shock-and-awe campaigns for which the general public has little appetite; and sometimes a quest for an ultimately unachievable vegan purity tends to cause skirmishes among the devotees. More recently, vegan critics have pointed out issues related to white supremacy in the vegan movement, and in particular its tendency to overlook or exclude Black, Brown, and Indigenous cultures' veganism, which was practiced as a form of resistance to colonialism (see Betty, 2021; Kennedy, 2021). There have been, of course, many groups elsewhere in the world, outside of mainstream Western veganism, who traditionally practiced vegetarianism or veganism for religious reasons; many of them are located in the Indian subcontinent, such as the Jains, Bishnoi, or the Todas (Tobias, 1996); also, vegan staples such as tofu and "mock meat" (which is today sometimes ridiculed as an "unnatural" alternative protein) have a rich and centuries-old history in Buddhist China where vegetarianism flourished. In short, there are multiple, intertwining veg(etari)anisms, with multiple points of origin and varied motivations and goals, and any attempt to contour their rhetoric must acknowledge this rich heterogeneity.

I should also clarify that during this chapter I alternate between vegetarian, veg(etari)an, vegetarian/vegan, and vegan for several reasons. The main term for abstention from meat has been, historically, "vegetarian" until the establishment of the Vegan Society in 1944 and the coining of the term "vegan" by Donald Watson and Dorothy Morgan to denote a broader abstention from consuming any and all animal products (including for clothing and other consumer goods). Since then, veganism has been acknowledged as the broader ethical practice and as a social justice movement, while vegetarianism has been sometimes used in parallel for describing strictly the dietary practice of abstaining from all animal products. Since the focus of this volume is primarily on the dietary practice, both vegetarianism and veganism can be used interchangeably in most cases, but we need to recognize that the modern term "veganism" denotes a much broader goal of eliminating all animal exploitation.

Understanding the motivations, purposes, and consequences of the vegetarian/vegan movements has been the focus of numerous investigations by historians, sociologists, anthropologists, psychologists, theologians, geographers, philosophers, and medical researchers, among others. These motives have historically ranged from religious and mystical to ethical and political to health-related ideologies. Hinduism, Buddhism, and Jainism practiced vegetarianism based on their reverence of life rooted in the doctrine of ahimsa, non-violence, or the doctrine of the reincarnation of souls. Pythagoras and his followers refrained from eating meat and animal products because they also believed in the existence of a transcendent soul that could be transferred into all other living beings; eating flesh, therefore, would have implied killing your kin. Various Christian sects well into the 19th and 20th century practiced vegetarianism also based on moral and ethical principles, as well as on claims of multiple health benefits; and 19th century vegetarians promoted the diet for moral and civic reasons (see Kondrlik, this volume). Nation-building politics sometimes relied on vegetarianism for questionable ideological reasons (such as in early 20th century Italy and Germany; see Treitel, 2017; Helstolsky, 2021). The second half of the 20th century sees the rise of intensive factory farms, a larger, ever-expanding scale of animal experimentation for scientific and medical purposes, and a rise in so-called lifestyle and chronic diseases—many of them diet-induced; consequently, many arguments for vegetarianism and veganism have targeted these perceived ills of late capitalist societies from an ethical, political, and medical perspective. Finally, as the consensus around human-induced climate change crystallizes early in the 21st century and identifies demand for meat and seafood as one of the main contributors to greenhouse gas emissions, deforestation, and earth and ocean pollution, environmental arguments become crucial in vegan rhetoric (see also Feltz and Arduser, this volume).

Diet is only one of the components of veganism, but a crucial one, since eating is a fundamental daily activity for all people on earth—unlike, say, choosing to wear leather or to go to an animal-based circus. As many food scholars and historians have pointed out, food and the many rituals associated with it are a source of identity, joy, and pride, a way to belong and to share in the community, a parameter for organizing time and labor, a paramount component of tradition, and a key element in health and healing practices (Fernández-Armesto, 2002). Furthermore, what we eat is both personal and political, as states have traditionally prescribed, directly or indirectly, dietary guidelines—either through "food pyramids" or similar recommendations, or through agricultural policies, subsidies, and other interventions. The persistence and ubiquity of meat in our diets has been dubbed carnism by Melanie Joy, which she defines as an ideology so pervasive that, like patriarchy or whiteness, it often goes unnoticed or is normalized and naturalized as simply the way things are (Joy, 2020). When so many of the traditions that contribute to our identity as members of a family, and so many of the external constraints that shape our behavior as a larger community are associated with meat and other animal products, what arguments can be deployed to rend those bonds, and overhaul not simply the content of meals but its far-reaching affective and political tendrils?

Veg(etari)an practice has almost always been animated by anti-establishment ethos; in fact, Potts and Armstrong call it "a form of resistant biopolitcs" (2018, 396). Thus, not surprisingly, abstention from meat has usually been regarded as fringe, abnormal, or deviant (Boyle, 2011; Taylor, 2012), especially in meat-centric populations. To combat this image of deviance and promote their message, vegetarians and vegans have had to resort to a vast array of rhetorical strategies targeting various motivations for meat consumption and animal use and abuse in general. These strategies, I argue, can be understood as a polyphonic or chameleonic rhetoric; such a framing may also offer a coherent way of understanding an otherwise decidedly non-monolithic and often fractured vegan/vegetarian movement that has seen its share of internal antagonism.

What Is a Polyphonic Rhetoric of Veg(etari)anism?

I borrow the term polyphony from Mikhail Bakhtin, who used to describe Dostoevsky's novels. For Bakhtin, Dostoyevsky was the first writer to truly create a polyphonic novel by creating unique characters and points of view in dialogue with each other but without imposing one single moralizing discourse or interpretation. Unlike homophonic narratives, in which the author's point of view is easy to detect and usually is subsumed to a specific and identifiably ideology, polyphonic novels de-center the omniscient subject and offer instead a more ambiguous, nuanced, dissonant conversation among multiple subjects whose perspectives matter equally: "a plurality of independent of unmerged voices and consciousnesses, a genuine polyphony of fully valid voices" (Bakhtin, 51). Bakhtin uses the term "dialogic opposition" (Bakhtin, 87) to describe Dostoevsky's technique—and perhaps philosophy of life: everything in life finds itself in such a dialogue. The very messiness and richness of life prevent the artist from imposing a unifying, flattening interpretation over them; rather, Bakhtin writes, "the artistic will of polyphony is a will to combine many wills, a will to the event" (Bakhtin, 66).

When looking at the contemporary landscape of vegetarian/vegan rhetoric today, it is therefore helpful to acknowledge that while the end goal might be similar (abstention from meat), the motivations and strategies used to persuade the public vary widely, and are also constructed in *dialogic opposition* to arguments used by meat eaters and promoters. A convenient way to summarize carnivore motives is offered by a 2015 study that identified the main reasons why people justify meat eating, which they summarized as the 4Ns: Natural, Necessary, Normal, and Nice (Piazza et al., 2015). The "Natural" category contained "appeals to biology, biological hierarchy, natural selection, human, evolution, or the naturalness of eating meat"; "Necessary" referred to "the necessity of meat for survival, strength, development, health, animal population control, or economic, stability;" "Normal" included "[a]ppeals to dominant societal norms, normative behavior, historical human behavior, or socially constructed food pyramids," and finally "Nice" referred to "appeals to the tastiness of meat, or that it is fulfilling or satisfying" (Piazza et al., 2015, 116). Additional categories listed respondents' beliefs that meat eating is justified because animals can be raised sustainably or slaughtered humanely, that their religion requires it, etc. Most of these arguments are not merely "individual choices" but internalized propaganda by the meat industry, as painstakingly documented by the collection *Meatsplaining: The Animal Agriculture Industry and the Rhetoric of Denial* (Hannan, 2020). Precisely because they are so pervasive, all of these "points" must and do have their dialogic counterpoints in vegan rhetoric.

Veg(etari)an Polyphony: An Attempt at Categorization

There are several strands of rhetorical vegetarianism/veganism, which I am loosely grouping under four categories: Ethical Idealism, Social and Ecological Justice, Medical and Cosmetic Neoliberalism, and Pragmatism. Each of these strands of rhetoric addresses the "4Ns" of eating meat in different ways. The remainder of the chapter is devoted to describing these types of veg(etari)an rhetorical engagement with the obvious caveat that each of them would merit a much lengthier discussion.

Ethical Idealism

Vegan rhetoric fueled by ethical idealism is animated by animal liberation as outlined in Peter Singer's landmark 1974 book; perhaps no group is more symbolic of this type of rhetoric than PETA, whose founder, Ingrid

Newkirk, was moved to action after reading Singer's book. Vegan rhetoric stemming from these ideals recognizes animals as fellow earthlings capable of sentience—including emotions and intelligence, and, most importantly, of feeling pain. Taking a cue from Jeremy Bentham's famous passage, "The question is not, Can they reason? nor Can they talk? but, Can they suffer?" ethical vegans reject any practice that exploits nonhuman animals, including animal testing, animal farming, animal "blood" sports like dog or bull fighting, animal entertainment (including circuses and zoos), wearing clothing and accessories made of fur, wool, leather, silk, bone, and other animal products, and eating anything of animal provenance. Gary Francione's total abolition program is a good example of an ethical idealist agenda (www.abolitionistapproach.com).

Often, pro-vegan arguments aim to overwhelm carnism with pathos, in hopes that building empathy for nonhuman animals may overcome the emotional attachment to meat. Vegan arguments from pathos urge us to consider the shared animality and sentience between human and nonhuman species. Advances in biological sciences have shown us that all animals, including the ones we most use in intensive animal farming operations, such as cows, pigs, and chicken, are capable of thinking, feeling, and suffering; this is also true for fish and seafood (though, to a certain extent, perhaps not bivalves; on that point the science is less clear). Raising these animals on the scale we do now involves well-documented methods that are, by any measure, cruel, barbaric, unsanitary, and exploitative of both human and nonhuman animals. And the production of meat is only increasing: in 2018, 80 billion animals were slaughtered for meat worldwide, and the trend is going up (Ritchie and Roser, 2019). The scale of animal suffering alone is staggering; from the utilitarian moral perspective championed by Singer, it is our duty to reduce that suffering.

PETA vegan campaigns often deploying a pathos-based rhetoric of discomfort, shock, and disgust, such as showing images of bloodied or tortured animal corpses as the reality behind one's meal—thus rendering visible what Carol Adams had famously dubbed as "the absent referent" (2016). Indeed, recent investigations into animal farming operations have uncovered horrific practices considered standard in the industry and which inflict untold suffering on farm animals, for example, gestation crates for sows, dehorning and branding cows, debeaking chickens, castrating pigs without anesthesia, crushing male chicks considered useless in the egg industry, separating newly born calves from their mothers and starving them in order to collect milk from cows, artificial insemination and frequent forced pregnancies, barbaric slaughter methods, neglect, overcrowding, unnatural diets, antibiotic overuse, runoff pollution, and so on. Such videos are often posted on websites and social media; they are so upsetting that Instagram, Facebook, or Twitter will mark them as "sensitive" and censor their previews in order to protect unwilling viewers: one needs consent and an extra click in order to reveal them. By contrast, documentaries like *Earthlings* or *Dominion*, which show footage of undercover investigations of concentrated animal feeding operations (CAFOs), require intentional viewing rather than automated feeds or social media scrolling; the visual impact of their findings is enormous, and difficult to neglect. Further PETA tactics involve "shockvertising" events in which, as one example, naked people[2] pose as lifeless bloodied cadavers in shrink-wrapped trays—there, the semiotic equivalency of human and animal meat plays into the "meat is murder" message. Similar campaigns have been deployed by animal liberation groups elsewhere (Véron, 2016).

To better understand the vegan argument from pathos, we can deploy Toulmin's famous argument structure (1958/2003). Toulmin posits that any argument consists of a claim backed by grounds, which are held together by a warrant, the often-implicit assumption that allows us to connect the claim with the grounds. The warrant can be further strengthened by backing, and the claim can always be further modified by possible rebuttals. In the case of arguments for meatlessness that offer the reality of factory farming as grounds for veganism, the warrant is that inflicting suffering on our fellow creatures is morally wrong. We should not eat meat because means of obtaining meat are cruel, and our common humanity should prevent us from committing cruelty, and—crucially—*supporting* it.

As the target audience, we would have to accept the warrant that there is little difference between our suffering and animal suffering, and therefore be moved by empathy. Furthermore, we would have to accept that we are in fact guilty by association whenever we pay and consume such products; as meat eaters, we have blood on our hands. This is where the argument has limitations beyond a select few, as many people, while ready to accept that factory farming is cruel, are not ready to accept they are responsible for it, or that human and animal suffering are equivalent, since animal lives would be less valuable in their view and subordinate to human needs. The roots of that belief matrix run deep—some would place them in the Genesis book of the Bible, where God confers "dominion" to man (sic) over sea and land creatures. Thus, veganism asks us to confront at a minimum two deeply ingrained belief systems around which we forge our identity: food and religion, both related to community and belonging; challenging them requires a lot of hard work and the reshaping of one's core values. It is, instead, easier than to reframe the ethical vegan as a deviant outcast (Boyle, 2011) and the "killjoy" at the table (Twine, 2014). Even though there is evidence that many of us do care about the welfare of both companion and farming animals, at least according to a recent survey of people in fourteen countries (Sinclair et al., 2022), our care is limited by perceived threats to our psychological and physical comfort. Bracketing the violence inherent in the production of meat, dairy, and eggs is a sine qua non condition of our ability to enjoy these products.

On the opposite spectrum of shocking/discomforting vegan arguments that tap into our guilt are appeals to our sense of comfort and savior instincts. Such arguments rely on showing us happy farm animals and asking to us to "go vegan—for them." Found both in PETA ads and in vegan social media posts by vegan activists like ChooseVeg, 1 Million Vegans, World of Vegans, and many others, these arguments from pathos (sometimes enhanced by the use of images of baby animals) appeal to our sense of compassion but take away the gore and create a niche that lets us imagine ourselves as heroes rather than as villains. Like many other digital support animals (which is what I call a booming social media genre that invites us to look at and take comfort in images of cute or funny animals, mostly pets but not only), images of happy farm animals outside the factory farm environment reinforce the ultimate goal of ethical veganism—animal liberation. Thus, the rhetoric of comfort complements its opposite, the rhetoric of shock and discomfort, by appealing to a different set of emotions and values, which are harder to resist or less likely to be critically examined by meat eaters than guilt-based discourse.

Ethical idealist vegans have often been criticized for their "radicalism"; we are urged instead to be "moderate" (a term that is poorly defined) and avoid "extremism." Of course, animals cannot be "moderately" dead for a "moderate" omnivorous diet; furthermore, for anyone who cares to investigate, industrial animal farming is a truly extreme practice. However, ethical idealists often veer into murkier waters when they condemn Indigenous tribes who rely on animals in their diet (even though in the far northern populations, animals are sometimes the only source of sustenance)—which has been roundly critiqued as a form of white, colonialist veganism. Another critique of ethical idealist vegans is that they tend to promote a type of dietetic purity that is hard, if not impossible to achieve, and then turn on those who cannot uphold such purity; for example, vegan food and culture writer Alicia Kennedy was publicly shamed (on Twitter and other forums) when she revealed that she started eating oysters in the wake of the sudden death of her brother, who was fond of this food. In reality, achieving a diet free of animal and human suffering is more of an ideal than a reality; for example, that staple of vegan diet, the banana, has a sordid past and an ethically ambivalent present, continuing a legacy of exploitative, neocolonial, and environmentally harmful practices (see Garcia, this volume). Furthermore, most crops, from planting to harvesting to storing, involve, purposefully or inadvertently, animal suffering—either through the use of fertilizers based on bone and blood meal, or by killing small wildlife living in the fields (e.g., small rodents, insects, etc.). The fresh produce a vegan diet relies on is collected at the expense of poor, marginalized, and exploited workers (see Holmes, 2013, for a detailed ethnographic account of the structural violence involved in a food system reliant on migrant work). As Lori Gruen and Robert Jones aptly put it, there are many reasons why veganism can only be an "aspiration" (Gruen and Jones, 2016). Purity, including in one's diet, is unachievable and ultimately an impediment to rather than an ideal of a good life (see Shotwell, *Against Purity*, 2016). Insisting on such purity is probably one of the reasons why ethical vegans are often perceived negatively by the public—or, as Kennedy quipped, "No one likes vegans, except other vegans, though sometimes even that is debatable" (2021).

Ethical idealism is prevalent in most current day advertising and vegan social media, and both rhetorical appeals to discomfort and guilt and to comfort and ego are widely deployed. It is, for the most part, narrowly focused on animal liberation and committed to practices that minimize animal suffering. It also mostly targets the third N of the 4Ns of eating meat: normal. Ethical vegan arguments rely heavily on de-normalizing the eating of meat and inching forward to the type of society described in Simon Amstell's mockumentary *Carnage* (2017). Filmed from the point of view of 2067, a year in which humanity is struggling to come to terms with their carnist past, *Carnage* depicts a world where eating meat is abnormal and abhorrent, akin to other questionable behaviors we left in the dustbin of history. A series of introspective characters ask themselves how we could have ever condoned the bloody business eating animal products, and try to make sense of it with gentle humor; flashbacks to supermarkets of yore (our current day stores) recast them in a sinister light. The mockumentary promotes a sense of alienation from the embodied practice of eating animal products, which is one of the primary goals of vegetarian rhetoric from an ethical idealism perspective.

Social and Ecological Justice

To an extent, this line of vegan reasoning is an offshoot of ethical idealism, but it goes a little bit further than the goals of animal liberation by aligning them with broader social and ecological justice values and highlighting how animal exploitation perpetuates environmental destruction as well as class and racial inequalities. It recognizes that the roots of speciesist oppression are entangled with colonialism, racism, patriarchal, and ableist exploitation. It shows how unchecked capitalism and drive to maximize growth and profit have led to environmental destruction and the suffering of both human and nonhuman animals alike. It looks at factory farming and meat-based diets not only in terms of their impact on the animals exploited but also in terms of pollution, deforestation, greenhouse emissions, and exploitation of factory workers and marginalized groups (e.g., undocumented laborers and low wage workers). It targets the systems that enable and perpetuate such exploitation and emphasizes the kinship between oppressed and marginalized groups. This type of rhetoric is anticapitalist, intersectional, and cognizant of the limits of a narrow focus on speciesism; it consequently demands that we also address the multiple oppressive systems that make speciesism possible—racism, sexism, ableism, and unbridled capitalism among them.

Writing for *Current Affairs* (2022), Maria Bolotnikova comments on the rhetoric of agricultural magazines typical of a period of growth for animal farming in the United States; a typical quote ran, "Forget the pig is an animal—treat him just like a machine in a factory." The chilling commodification of animal bodies reminds her of chattel slavery; we in effect are urged to dissociate our affect and effectively equate sentient beings with production units, in the name of profit. This is true of animal and human capital alike. Astra and Sunaura Taylor put it bluntly: "Capitalism turns bodies into machines. Like their predecessors on the earliest factory lines, today's workers are compelled to perform like robots, whether packing shipments in Amazon's warehouses or driving for UPS or Uber. This process of mechanization and standardization affects not only the bodies of human laborers but also unwaged women, as well as cows, chickens, and pigs" (2022, para 8). Some sociologists argue that we have in fact started this process of commodification the moment we domesticated species and invented animal agriculture in order to use animal bodies and their labor; David Nibert argues that a more apt term for domestication is "domesecration":

> The widespread violence and destruction engendered by such uses of large numbers of "domesticated" animals encompasses both the violence experienced by the animals and the ways in which this harm has been entangled with related forms of violence against free-living animals and groups of devalued humans. These include invasion, conquest, extermination, displacement, repression, coerced and enslaved servitude, gender subordination and sexual exploitation, and hunger. Accompanying such violence have been deadly zoonotic diseases that have contributed to the destruction of entire cities, societies, and civilizations."
>
> (2013, 5)

Recognizing the role animal exploitation and consumption have played in the Anthropocene, including the extinction of countless species,[3] is a first step to fighting a system that has pushed more and more unsustainable practices effectively amount to what vegan anarchist geographer Simon Springer calls "ecocide." Springer argues for veganism (or rather, veganarchism) as part of a policy of "total liberation" for human and nonhuman animals as an antidote to our current ecocidal practices. For him, veganism "… is an ethical response to the senseless and unnecessary mass murder of sentient beings caught up in the violent machinery of contemporary capitalism. It is a commitment to the defiance of human supremacy, which represents the same reprehensible scourge as all forms of supremacy, whether in support of race, gender, sexuality, age, or any other identity category. The implication of such an integral approach to violence is that none are free until all are free" (2021, 249). Vegan geographers, bioethicists, and environmental scholars also urge us to consider the implications of animal farming for global health, as meat production has been convincingly linked to epidemics, including COVID-19 (Vettese and Blanchette, 2020). Vettese and Pendergrass have a bold proposal for a postcapitalist society that will stem global environmental destruction, one that includes veganism as a necessary step (*Half-Earth Socialism*, 2022).

Carol Adams has famously linked eating meat, dairy, and eggs to the sexual exploitation of women, painstakingly deconstructing the role of meat as a symbol of masculinity in *The Sexual Politics of Meat* (1990/2016). In this volume, Abby Dubisar and Erin Trauth take on this conundrum and the role masculinity plays out in the construction of the vegetarian burger. Ecofeminism (as exemplified by the work of Greta Gaard, Val Plumwood, and Marti Kheel) is also animated by systemic intersectional justice although it balances it against the interests of the land and the people on it, which may not always be strictly vegan—as in the case of subsistence farming and animal diets for indigenous people; see also Gruber, this volume. Laura Wright in *The Vegan Studies Project* (2015) further develops the project of ecofeminism and uses veganism as a framework through which we can interpret cultural phenomena and become more sensitized to the plight of vulnerable populations.

A. Breeze Harper, author of the *Sistah Vegan Project*, cites Marjorie's Spiegel *The Dreaded Comparison* and Charles Patterson's *Eternal Treblinka* among the major influences leading to her adoption of an ahimsa-based veganism. The comparison between the plight of animals and the plight of enslaved Black people or of Jewish people in the Holocaust has also been used by PETA and similar groups, but Harper argues that this was done insensitively and disrespectfully (mostly through "shockvertising"); instead, she urges us to recognize "the interconnectedness of institutionalized racism, nationalism, and sexism; the mistreatment of nonhuman animals; and the abuse of the planet's natural resources" (2010, ix). Lisa Betty argues that mainstream veganism, at least in the United States, is mainly white and dominated by a white supremacist ideology, and it has been so by design, as it failed to recognize its wide liberatory potential (2021). She also urges us to contend with the problematic neocolonial aspects of veganism especially when they promote crops that have only been possible through the enslavement or exploitation of "Black and brown indigenous peoples throughout the Global South tropical zones all in the name of coffee, cocoa, coconut, avocado, almond, and banana" (2021, para. 27) (see also Garcia, this volume). Sunaura Taylor's groundbreaking *Beasts of Burden* (2017) draws parallels between her own disability and animal disability, and between her treatment by society and animals' treatment in society. An awkward encounter described in the book between the author and Peter Singer, who has famously been criticized for his ableist positions, serves as a further illustration of the distance between ethical veganism and veganism animated by global social and ecological justice.[4]

A vegan rhetoric rooted in social justice and ecological ethics is a type of *deliberative* rhetoric; it is partially rooted in logos, as it connects history, politics, and environmental science, and partly in pathos, as it essentially appeals to our empathy and survival instincts. It furthers the project of ethical veganism and asks us to recognize the kinship between all species, human and nonhuman alike, and to rally against injustice against all species and ecosystems that support them. Vegan rhetoric rooted in social and ecological justice is a political project that targets the first N of the 4Ns of eating meat: natural. There is, in fact, nothing natural about eating meat—moreover, it is an act that is deeply destructive to the environment and life on earth as we know it.

Medical and Cosmetic Neoliberalism

One of the main ways in which the public is introduced to veganism is what I call medical neoliberalism: the notion that we are responsible for our health through our lifestyle, and therefore must adopt the ideal diet to optimize our health outcomes (this has also been called healthism—a term coined by Richard Crawford in 1980). Veganism is in this context prescribed as a cure to chronic lifestyle diseases on the rise in Western countries (hypertension, heart disease, diabetes, and so on), and generally presented as capable of preventing such ills. Doctors like T. Colin Campbell (author of *The China Study*) and Neal Barnard (author of numerous books about the benefits of a vegan diet) are some of the best-known medical authorities promoting veganism as a preventative and curative regimen (see also D. R. Hammontree, in this volume, on how the ethos of such vegan medical celebrities is built and contested). Many public figures went vegan for health reasons—including former president Bill Clinton (he is now known as a "chegan," a cheating vegan occasionally eating fish and eggs). This is an aspect of vegetarianism that holds endless appeal in the media economy, though it makes little sense for ethical vegans; to quote Isaac Bashevis Singer when he was asked whether he was vegetarian for health reasons: "Sure, for the health of the chicken."

Additionally, the neoliberal reasons for adopting a vegan diet often conflate health with thinness. There is a particular strand of "cosmetic veganism" that promotes it as a way to stay fit, thin, and young; some examples are the popular diet book *Skinny Bitch* (written by a former model and her agent) or *Ageless Vegan*, by Tracye McQuirter. One time YouTube phenomenon Leanne Ratcliffe, aka Freelee the Banana Girl (whose channel still has 800,000 subscribers, and who once boasted up to 10 million daily views) promoted a high-carb low-fat vegan diet *primarily* for weight loss (this involved eating no less than thirty bananas a day). None other than Beyoncé Knowles adopted, on and off, a vegan diet for the same reason, even getting involved with a commercial program that encourages adopters to go vegan for 22 days. It is not hard to see how this sort of cosmetic framing of a vegan diet can be a slippery slope from the point of view of ethical or social justice vegans. In their analysis of the "Beyoncé diet," Fegitz and Pirani (2017) argue that it diminishes the meaning of self-empowerment and imposes a strict surveillance bodily regimen that is detrimental to feminist and social justice causes: "while eco-feminists have embraced vegetarian and vegan regimes as ethical and political choices, post-feminism depoliticizes and deradicalizes them" (1).

The alignment of veg(etari)anism with a variety of trends resulting from healthism, food-faddism, pseudoscience, and "wellness" broadly defined further complicates the deceptively "simple" choice to go veg(etari)an, and can serve to further undermine it, as the public appetite for lifestyle controversies is seemingly endless. Jordan Younger, formerly The Blonde Vegan, wrote a whole memoir about how she realized her veganism was an eating disorder (*Breaking Vegan*, 2015), and then rebranded herself as The Balanced Blonde. Stories like hers were heavily mediatized, not only because she was a popular influencer, a young and pretty white blonde woman in the public eye, and an eating disorder cautionary tale, but also because vegans are widely disliked and seeing them fail or revealing their veganism as a disease or health failure justifies that dislike (see also Wright, 2015, who discusses this trope at length). This is just one of the issues with framing veganism as a health imperative. There will always be counterstories that are persuasive because they rely on dramatic irony (peripeteia), which, in the attention economy, brings in large numbers of readers while feeding into the mainstream narrative of strident, unlikable, "killjoy" vegans. Such narratives also have more serious consequences, such as in stigmatizing and sometimes criminalizing vegan parenting based on the view that raising a child vegan constitutes criminal neglect.[5]

Essentially, a vegan rhetoric stemming from medical neoliberalism is based on logos: it relies on scientific arguments for the merits of a vegan diet, or—in the case of cosmetic view of veganism—on visual proof that a vegan diet produces thinness, beauty, and health (Freelee the Banana Girl or the *Skinny Bitch* authors are cases in point). It takes issues with the "necessary" part of the 4Ns of eating meat: it argues that there is nothing necessary about meat in our diets. Two further issues arise. The first is the cooption of veganism by wellness culture, which often veers into pseudoscience, and which has the potential to undermine on that count alone the practice of veganism by equating it with food fads and reducing it to a strictly consumer issue. By that logic, veganism is only practiced as long as it is proven useful to the individual consumer's physique, thus reinforcing a typical neoliberal solipsism. The second is the difficult issue of scientific consent. While many scientific papers show the dangers of diets high in animal products and tout the health benefits of a largely vegetarian diet, the case is still not clear that similar health results cannot be achieved by incorporating a *small* amount of animal products in one's diet; for example, the Mediterranean diet, widely studied and recognized as one of the healthiest diets in the world, incorporates moderate quantities of fish and other animal products. There is also debate in the scientific community over optimal protein quality (animal or plant), and micronutrient importance (iron and B12 are among the micronutrients that are more readily available in animal rather than plant products). Scientifically speaking, there is no true consensus that a vegan diet is optimal—just that it is *healthier* than one heavy in animal products; conversely, we cannot prove that incorporating *modest* amounts of animal protein in our diet is decidedly unhealthy. Additionally, plenty of vegan food can be unhealthy, too, especially with the proliferation of vegan fast and junk food options.

Thus, based on logos alone, the pro-vegan argument from health or cosmesis may have several inconsistencies, or can be easily attacked; from a lay public's perspective, the minutiae of nutrition research seem arcane, and what transpires is, at best, a firm case for reducetarianism (reducing consumption of animal foods) rather than for full veganism. The research seems to preach to the already converted—it may both convince omnivores that moderation is the way to go, and vegans that their diets are the healthiest as long as they supplement it with B12 vitamins. Consequently, a vegan rhetoric based on medical neoliberalism and cosmetic benefits veers the furthest away from the aims of a vegan practice, by centering itself on an illusory scientific (and sometimes pseudoscientific) consensus and by entirely bracketing any ethical or broader intersectional and environmental justice aims. However, even though the grounds are contested, a health-based vegan rhetoric comes with a strong warrant that we are conditioned to accept under neoliberalism, namely, that we are individually responsible for our health and need to control it through our diets. This warrant, though sometimes problematic, still relies on the fact that diets rich in plant products are indisputably healthier, and it has both the public's and the policy makers' attention. Whereas current agricultural and economic policies sanctioned by the state make it very difficult for vulnerable populations to access affordable fresh fruits and vegetables and quality plant-based protein,[6] further vegan advocacy from a health angle may help move the public opinion needle and eventually help us embrace better, more sustainable, more humane, and more rational food policies.

Pragmatism

A pragmatic approach to vegetarianism/veganism recognizes that selling an agenda of ethical abstention from meat and/or of total animal liberation to the public is a high order, perhaps an impossible one, not only conceptually, but, more importantly, practically. The warrant that ethical idealistic arguments ask meat eaters to accept—the recognition that the suffering of our fellow species is no less than our own, that condoning that sort of suffering is morally wrong—is often forcefully rejected for varied reasons—ranging from religious to lip service to "humane" farming and slaughter, to taste preferences, or appeal to tradition (Thanksgiving turkeys, Christmas hams, Fourth of July hot dogs, grandma's cooking, and everything in between). The warrant that asks us to recognize intersectional and ecological justice is, similarly, a bridge too far for many. The neoliberal warrant of individual responsibility for one's health comes with its own set of problems, as outlined above. Once these warrants are denied, the argument(s) for veganism breaks down, and new warrants need to be pursued. Pragmatists recognize the nearly insurmountable difficulties presented by the heterogeneity of motives for meat consumption, and propose an incremental approach that would reduce, rather than eliminate animal consumption, based on a shared assumption that moderation is desirable (therefore, eating less meat makes reasonable sense) and that meat equivalents are equally nutritious or tasty. Such moderate utilitarian solutions are reducetarianism or flexitarianism (e.g., Meatless Mondays, or Mark Bittman's VB6/Vegan before 6 program, or rebranding veganism as "plant-based lifestyle"), vegan hedonism (which proposes rich, lush vegan recipes to replace meat-centric ones), and technological veganism, which proposes "alt proteins" or even cellular agriculture (cultivated meat) as meat replacements.

These strands of pragmatic vegan rhetoric are often animated by the same noble ideas as ethical idealism or social and ecological justice, but offer different solutions—after all, what is more likely to happen, a complete overhaul of capitalist principles on a global scale, or that some people will sometimes replace their cow burger with a Beyond Beef one? In the public arena, when these arguments play out, there is no reference to Bentham's "Can they suffer?" or to intersectional justice. There is one exception: as the reality of climate change becomes harder to ignore, the arguments for meat reduction and plant-based diets grow stronger. This warrant ("we all care for the environment, and eating less meat helps protect it") is often referenced in pragmatic vegan arguments. Such appeals have been helped along by the evolution of public discourse around climate change, and by popular documentaries such as *Cowspiracy*, which also took animal suffering out of the equation and focused instead on methane emissions and other types of pollution produced by cattle operations. The younger generations seem to be particularly receptive to such rhetoric (see Weik von Mossner, 2020).

For example, the creators behind Bosh TV, a vegan online cooking channel, offer three reasons for their enterprise: "We live on the Veg," "We love food," and "We dig the planet," which they further explain as follows: "The production of meat requires huge outlays of pesticides, fertiliser, fuel and water while releasing greenhouse gases. Eating more plants is a small change that makes a big ol' impact" (bosh.tv, 2022). There are zero moral references. Bosh TV and the many other sensual vegan foodies of Instagram, YouTube, and TikTok are offering compressed lush videos documenting decadent recipes that range from high end 50-ingredient salads to vegan junk food. While ethical motives can be implied, the rhetorical strategies used in this sort of vegan outreach could not be more different. The argument is based on equivalency: if all else is equal (e.g., vegan food is just as tasty as traditional/animal-based food), then why not choose the diet that offers extra benefits such as reducing greenhouse emissions? The vegan foodie angle is a pragmatic one because it offers alternatives rather than moralistic pronouncements, and foregrounds taste over guilt trips; where an ethical idealistic argument would prompt us to ask if we value our taste more than the life of sentient beings, vegan foodies show us lush, exuberant, mouthwatering food that happens to be vegan. Vegan hedonism attacks the "nice" of the "4Ns" of eating meat, or rather, unsettles it: eating meat alternatives is just as nice, if not nicer!

Jan Dutkiewicz and his collaborators (Simon Rosenberger, Matthew Hayek, Jonathan Dickstein, and others) are firmly in the pragmatist camp. Dutkiewicz and Dickstein in fact argued for an explicit redefinition of vegan as strictly of practice of abstention from meat, and in favor of stripping it from ethical and other intersectional justice connotations (2021). These activist, ethical, or otherwise charged connotations, they argue, are confusing, since vegan motivations are vastly heterogeneous. Dutkiewicz and Dickstein, in other words, dislike the rhetorical polyphony of the vegan movement and see it as infusing veganism with "incoherence as a belief-ism" (2021, 11). Applying the broad umbrella of veganism to a variety of fundamentally different ideological movements puts the cart before the horse—veganism is a product of many diverse strands of thought and motivations, and it is both unfair and incoherent to critique veganism as a practice because of perceived issues with any of these motivations. Not surprisingly, Dutkiewicz's public and academic work has focused lately on alternative (plant) proteins and cellular agriculture as viable options to meat production, which have the potential of vastly reducing animal suffering by cutting down (and eventually eliminating) the need for CAFOs, with the side effect of being more beneficial for the environment.

One of the dangers, however, of pushing such technologies is that they can be coopted, de-radicalized, and used to "greenwash" corporations built on animal exploitation; giant chicken corporations like Tyson and Purdue have invested heavily in alternative proteins, for example, without really scaling down their chicken production. The rhetoric of selling such products is explored by Buscemi in this volume. Furthermore, biotechnological solutions offering "meat" substitutes come with their own set of problems—such as the assumption that meat protein is a necessary requirement of a balanced diet (and therefore we must find perfect plant-based equivalents), or the assumption that one needs to experience "meatiness" in plant-products by any means necessary, such as making them "bleed" (see Trauth, this volume). Another issue is the ultra-processed nature of such products, which is pitted against the "organic" and "natural" labels usually associated with veg(etari)anism.[7] Cellular agriculture or in vitro meat, at least as of now, is not meat or suffering free, as it uses stem cells from calf fetuses, nor environmentally friendly, as it requires vast amounts of energy compared to its protein output; Nathan Poirier has in fact argued that it constitutes "the epitome of (industrial) animal agriculture in terms of history, material practices, and ideology" (2022, 1). Technological progress that mitigates some of those ethical and environmental concerns may be fast on this front, but it may not keep pace with the accelerated rhythm of climate change. In other words, we know that industrial farming is a large contributor to climate change; while presenting people with viable alternatives for meat production that does not rely on industrial farming is a noble goal and might help us mitigate climate change, it is not happening fast enough or on the scale necessary for its effects to take place. Higher level policy changes may help us along, such as reducing or eliminating subsidies for meat production, or supporting the research and production of alternative protein; and indeed, pragmatic veganism is nothing if not deliberative, offering practical solutions for our immediate future. So far, however, we are nowhere near where we would need to be on this front.

Pragmatic vegans bet on the public's willingness to adopt rational solutions to today's crises—foremost among them, climate change;[8] they aim to make meat abstention *moderate*, thus removing the stigma of deviance from it. While this is certainly a worthwhile goal, as a pragmatic rhetorician, I am skeptical that stripping "veganism" of all its ideological connotations is going to accomplish that. I am, of course, willing and hopeful that I will be corrected. However, most people's attachment to meat is not purely pragmatic, but deeply affective. Meat-centric food traditions and social rituals are both embodied and embedded in affective memory; as such, they are hard to pry off with an intellectual wedge. Furthermore, meat eaters also define themselves along multiple lines of reasoning, which may indicate that they may be persuaded to consider a vegan practice by a variety of arguments.

While Dutkiewicz and Dickstein's proposal offers a sort of Occam's Razor simplicity, it is hard to see it pan out for another simple reason: vegetarianism has been from the beginning a value-based decision. In other words, one's core values have always, already been part of it—whether they be reverence for all life, justice for animals, or environmental or health beliefs. Dutkiewicz's campaign for alt proteins and cell ag is propped up by reasonable ecological concerns as well as by an underlying technophilia. In the public arena, arguments for veg(etari)anism are never presented without attachment to such beliefs—without us being asked to accept *new* warrants that speak to our core beliefs. Pragmatic vegan rhetoric de-centers warrants based on animal suffering and intersectional justice and asks us to make the most rational choice that will minimize harm to the planet and to ourselves while at the same time not sacrificing taste and tradition. One wouldn't be wrong to see this is as a very tall order.

Conclusion

Veganism is an embodied practice that can be the logical result of a multitude of disparate beliefs; it is, consequently, never a neutral or "simple" practice of abstention. Consequently, the rhetorical strategies used by the many flavors of vegan movements are polyphonic and, sometimes contradictory.

Let me reprise here Bakhtin's definition of polyphony: "a plurality of independent of unmerged voices and consciousnesses, a genuine polyphony of fully valid voices." The four major categories of vegan rhetoric identified, however imperfectly, in this chapter, are all valid voices entering in "dialogic opposition" with one another as well as with meat-centric rhetorical arguments, which are synthesized by the "4Ns" of eating meat.[9] All of these types of vegan rhetoric ultimately help to move us closer to a vegan future—one that is profoundly necessary for our planetary survival, that is fair and ethical toward our fellow species, and that is overall healthier for us as individuals. A polyphonic vegan rhetoric thus combines, to use again Bakhtin's words, "many wills, a will to the event": a widespread vegan practice. Such a practice may be, in fact, just the beginning of learning a new way to exist in the world without inflicting further harm on human and nonhuman animals—this includes eating locally and seasonally, and finding ways of growing and harvesting food that do not deplete the soil, displace species, create poisonous legacies, or exploit humans (see Betty, 2021; Garcia, this volume).

The current polyphonic rhetoric of veganism, albeit "messy" and contradictory at times, offers multiple avenues toward a vegan future. Even though disagreements and skirmishes among vegan groups may indicate vulnerability and may be exploited by outsiders as anti-vegan points, a polyphonic understanding of these arguments lets us see them as internally coherent, valid on their own, appealing to diverse audiences, and engaging in dialog with carnism on multiple fronts. All of them (including the medical/cosmetic neoliberal strand of rhetoric) have great potential to produce not just debates but political action as they move the public closer to accepting the widespread benefits of veganism. As Potts and Armstrong argue, a vegan world aims to reduce misery and create more joy (2018, 407). Such a world will satisfy the utilitarian imperative for reducing suffering, will work to mitigate global justice for all species, may make us healthier, and will help mitigate global change. Most urgently, importantly, when we know the world is on fire, we cannot fuel the flames by mindlessly endorsing the intensive commodification and consumption of animals: for that, we must use all the fabled Aristotelian "available means of persuasion"—vegan rhetoric in all its polyphonic glory.

Notes

1 I own a debt of gratitude to Vasile Stanescu, whose passionate talk "The World Is on Fire" (delivered in a webinar organized by Corey Wrenn at the University of Kent, 2020) inspired some of the points in this chapter.

2 The use of naked models (including celebrities) by PETA has been frequently criticized as sexist—see Deckha (2008), Glasser (2011) among others.

3 Coimbra et al. (2020) argue that human carnivory is a major driver of vertebrate extinction and our current biodiversity crisis through several mechanisms among which are predation, competition, biohazards, and environmental changes.

4 Singer cannot fathom that Taylor views her disability not only as acceptable, but as desirable, which goes contrary to his utilitarian perspective of minimizing suffering. In doing so, he comes close to denying Taylor's humanity and right to exist.
5 There are many instances of such cases around the world; even in cases where the parent was clearly guilty of neglect, the insertion of the descriptor "vegan" casts definitive aspersions and assumptions of guilt by association, as in the 2022 murder trial of Florida mother Sheila O'Leary (www.lawandcrime.com), reported widely as "Vegan mom convicted" (rather than simply "Mom"). Cases of parental neglect are not altogether uncommon, but where the descriptor "vegan" is involved, veganism is villainized by default, along with the parent.
6 See Guthman (2011) and Lozon (2021).
7 Much more could be said here about the anti-industrial sentiment of vegan subgroups and the interpretation of "natural"—which is not so much a descriptive as a quasi-religious term, as Alan Levinovitz brilliantly argues in his book *Natural: How Faith in Nature's Goodness Leads to Harmful Fads, Unjust Laws, and Flawed Science* (2020).
8 As history has taught us over and over again, this is a risky proposition. A recent case in point: choosing to wear a mask or not at the height of the COVID epidemic, 2020–2022. Ideology often won over rationality in that case.
9 These are, of course, a simplification of the totality of methods for legitimizing meat, albeit a useful one.

Bibliography

Adams, Carol J. *The Sexual Politics of Meat – 25th Anniversary Edition: A Feminist-Vegetarian Critical Theory*. New York, London: Bloomsbury Academic, 2016.

Amstell, Simon, Director. *Carnage* [Film]. *BBC*, 2017.

Bakhtin, Mikhail. *Problems of Dostoevsky's Poetics*, edited by Caryl Emerson. Minneapolis, MN: University of Minnesota Press, 1984.

Betty, Lisa. "Veganism* Is in Crisis. *As an Anti-oppressive Social (Justice) Movement." *Medium*, 2021. https://lbetty1.medium.com/veganism-is-in-crisis-36f78fa9a4b9 Accessed 10 July, 2022.

Bolotnikova, Maria. "Forget They Are an Animal." *Current Affairs*, August 1, 2022. https://www.currentaffairs.org/2022/08/forget-they-are-an-animal. Accessed August 20, 2022.

Boyle, Joseph E. "Becoming Vegetarian: The Eating Patterns and Accounts of Newly Practicing Vegetarians." *Food and Foodways*, vol. 19, 2011, pp. 314–333. DOI: 10.1080/07409710.2011.630620

Coimbra, Zulmira H., Luiz Gomes-Jr, Fernando A.S. Fernandez. "Human Carnivory as a Major Driver of Vertebrate Extinction." *Perspectives in Ecology and Conservation*, vol. 18, no. 4, 2020, pp. 283–293. DOI: 10.1016/j.pecon.2020.10.002

Crawford, Richard. "Healthism and the Medicalization of Everyday Life." *International Journal of Health Services*, vol. 10, no. 3, 1980, pp. 365–388.

Deckha, Maneesha. "Disturbing Images: PETA and the Feminist Ethics of Animal Advocacy." *Ethics and the Environment*, vol. 13, no. 2, 2008, pp. 35–76.

Dutkiewicz, Jan and Jonathan Dickstein. "The Ism in Veganism: The Case for a Minimal Practice-based Definition." *Food Ethics*, vol. 6, no. 2, 2021. DOI: 10.1007/s41055-020-00081-6

Fegitz, Ella and Daniela Pirani. "The Sexual Politics of Veggies: Beyoncé's "Commodity Veg*ism." *Feminist Media Studies*, vol. 18, no. 2, 2017, pp. 294–308. DOI: 10.1080/14680777.2017.1358200

Fernández-Armesto, Felipe. *Near a Thousand Tables: A History of Food.* New York, London, Toronto, Sydney: Free Press, 2002.

Food and Agriculture Organization of the United Nations [FAO]. *Tackling Climate Change through Livestock. A Global Assessment of Emissions and Mitigation Opportunities.* https://www.fao.org/news/story/en/item/197623/icode/. 2013. Accessed August 22, 2022.

Francione, Gary. *Animal Rights: The Abolitionist Approach.* https://www.abolitionistapproach.com/. Accessed August 20, 2022.

Freedman, Rory and Kim Barnouin. *Skinny Bitch: A No-nonsense Guide for Savvy Girls Who Want to Stop Eating Crap and Start Looking Fabulous!* Philadelphia and London: Running Press, 2005.

Gaard, Greta. "Vegetarian Ecofeminism: A Review Essay." *Frontiers: A Journal of Women Studies*, vol. 23, no. 3, 2002, pp. 117–146.

Glasser, Carol L. "Tied Oppressions: An Analysis of how Sexist Imagery Reinforces Speciesist Sentiment," *The Brock Review*, vol. 12, no. 1, 2011, pp. 51–68. DOI: 10.26522/br.v12i1.333

Gruen, Lori and Robert Jones. "Veganism as Aspiration." In *The Moral Complexities of Eating Meat*, edited by Ben Bramble and Bob Fisher. Oxford University Press, 2016, pp. 153–171.

Guthman, Julie. *Weighing in: Obesity, Food Justice, and the Limits of Capitalism.* Berkeley, CA: University of California Press, 2011.

Hannan, Jason. *Meatsplaining: The Animal Agriculture Industry and the Rhetoric of Denial.* Sydney: Sydney University Press, 2020. DOI: 10.30722/sup.9781743327104

Harper, A. Breeze. *Sistah Vegan: Black Female Vegans Speak on Food, Identity, Health, and Society.* Brooklyn, NY: Lantern Books, 2010.

Helstolsky, Carol. "State of Meatlessness: Voluntary and Involuntary Vegetarianism in Early Twentieth-Century Italy." In *Veg(etari)an Arguments in Culture, History, and Practice: The V Word*, edited by Cristina Hanganu-Bresch and Kristin Kondrlik. Palgrave/MacMillan, 2021, pp. 3–24.

Holmes, Steh M. *Fresh Fruit, Broken Bodies. Migrant Farmworkers in the United States.* Berkeley, CA: University of California Press, 2013.

Joy, Melanie. *Why We Love Dogs, Eat pigs, and Wear Cows: An Introduction to Carnism.* Newburyport, MA: Red Wheel, 2020.

Kennedy, Alicia. "On Veganism, and Why It Shouldn't Be Easily Mocked and Written Off." http://aliciakennedy.news (Newsletter). March 9, 2021.

Levinovitz, Alan. *Natural: How Faith in Nature's Goodness Leads to Harmful Fads, Unjust Laws, and Flawed Science.* Boston, MA: Beacon Press, 2020.

Lozon, Emma. "The Dialectic of Food Swamps and Clean Food: Ecological Interventions for Disrupting Individualizing Frames of Food Choice," *Rhetoric of Health and Medicine*, vol. 4, no. 2, 2021, pp. 218–245. DOI: 10.5744/rhm.2021.2005

Naham, Matt. "Vegan Mom Convicted of Murdering 18-Month-Old Son by 'Severe Malnutrition and Dehydration' Fails to Get New Trial." 15 August, 2022. https://lawandcrime.com/crime/vegan-mom-convicted-of-murdering-18-month-old-son-by-severe-malnutrition-and-dehydration-reportedly-fails-to-get-new-trial/. Accessed August 22, 2022.

Nibert, David. *Animal Oppression and Human Violence: Domesecration, Capitalism, and Global Conflict*. New York, NY: Columbia University Press, 2013.

Piazza, Jared, Matthew B Ruby, Steve Loughnan, et al. "Rationalizing Meat Consumption. The 4Ns." *Appetite* vol. 91, 2015, pp. 114–128. DOI: 10.1016/j.appet.2015.04.011.

Poirier, Nathan. "On the Intertwining of Cellular Agriculture and Animal Agriculture: History, Materiality, Ideology, and Collaboration." *Frontiers in Sustainable Food Systems*, vol. 6, 2022, p. 907621. DOI: 10.3389/fsufs.2022.907621.

Potts, Annie and Philip Armstrong. "Vegan." In *Critical Terms for Animal Studies*, edited by Lori Gruen. Chicago, IL: University of Chicago Press, 2018, pp. 394–407.

Ritchie, Hannah and Matt Roser. Meat and Dairy Production. *Our World in Data*. 2017/2019. https://ourworldindata.org/meat-production. Accessed August 12, 2022.

Shotwell, Alexis. *Against Purity: Living Ethically in Compromised Times*. Minneapolis, MN: University of Minnesota Press, 2016. Kindle edition.

Sinclair, Michelle, Lee Natasha Y. P., Hötzel Maria José, de Luna Maria Catalina T., Sharma Arvind, Idris Musadiq, Derkley Tessa, Li Congcong, Islam Mohammad Ariful, Iyasere Oluwaseun S., Navarro Grisel, Ahmed Abdelkareem A., Khruapradab Chanadda, Curry Michael, Burns Georgette Leah, Marchant Jeremy N. "International Perceptions of Animals and the Importance of Their Welfare." *Frontiers in Animal Science*, vol. 3, 2022. DOI: 10.3389/fanim.2022.960379

Springer, Simon. "Total Liberation Ecology: Integral, Anarchism, Anthroparchy, and the Violence of Indifference." In *Undoing Human Supremacy: Anarchist Political Ecology in the face of Anthroparchy*, edited by Simon Springer, Jennifer Mateer, Martin Locret-Collet, and Maleea Acker. London: Rowman & Littlefield, 2021, pp. 235–253.

Stanescu, Vasile. "The World Is on Fire: Animal Agriculture, Climate Change, and the Path Forward." Talk given at the University of Kent Faculty Seminar Series, March 25, 2021. https://www.kent.ac.uk/events/event/48013/the-world-is-on-fire-animal-agriculture-climate-change-and-the-path-forward

Taylor, Chlöe. "Abnormal Appetites: Foucault, Atwood, and the Normalization of an Animal-Based Diet." *Journal for Critical Animal Studies*, vol. 10, no. 4, 2011, pp. 130–148.

Taylor, Sunaura. *Beasts of Burden: Animal and Disability Liberation*. New York, NY: The New Press, 2017.

Taylor, Astra and Sunaura Taylor. "Our Animals, Ourselves: The Socialist Feminist Case for Animal Liberation." *Lux Magazine*, vol. 3, 2022. https://lux-magazine.com/article/our-animals-ourselves/

Tobias, Michael. "The Anthropology of Conscience." *Society and Animals*, vol. 4, no. 1, 1996, pp. 65–73.

Toulmin, Stephen. *The Uses of Argument*. Updated Edition. Cambridge, UK: Cambridge University Press, 2003.

Treitel, Corinna. *Eating Nature in Modern Germany. Food, Agriculture, and Environment, c. 1870–2000*. Cambridge, UK: Cambridge University Press, 2017.

Twine, Richard. "Vegan Killjoys at the Table—Contesting Happiness and Negotiating Relationships with Food Practices." *Societies*, vol. 4, 2014, pp. 623–639. DOI: 10.3390/soc4040623.

Véron, Ophélie. "(Extra)ordinary activism: veganism and the shaping of hemeratopias." *International Journal of Sociology and Social Policy*, vol. 36, no. 11/12, 2016, pp. 756–773.

Vettese, Troy and Alex Blanchette. "Covid-19 Shows Factory Food Production Is Dangerous for Animals and Humans Alike." *The Guardian*, September 8, 2020. https://www.theguardian.com/commentisfree/2020/sep/08/meat-production-animals-humans-covid-19-slaughterhouses-workers.

Vettese, Troy and Drew Pendergrass. *Half-Earth Socialism*. London: Verso, 2022.

Weik von Mossner, Alexa. "Feeling Bad? Veganism, Climate Change, and the Rhetoric of Cowspiracy. State of Meatlessness: Voluntary and Involuntary Vegetarianism in Early Twentieth-Century Italy." In *Veg(etari)an Arguments in Culture, History, and Practice: The V Word*, edited by Cristina Hanganu-Bresch and Kristin Kondrlik. Palgrave/MacMillan, 2020, pp. 245–269.

Wright, Laura. *The Vegan Studies Project: Food, Animals, and Gender in the Age of Terror*. Athens, GA: University of Georgia Press, 2015.

Younger, Jordan. *Breaking Vegan: One Woman's Journey from Veganism, Extreme Dieting, and Orthorexia to a More Balanced Life*. Beverly, MA: Quarto Publishing, 2015.

2 The Power of Ecological Rhetoric

Trans-Situational Approaches to Veganism, Vegetarianism, and Plant-Based Food Choices

Sibylle Gruber

An Introduction to the Complexity of Food Choices

In a 2017 report on three studies examining bias against vegetarians and vegans in the United States, Cara MacInnis and Gordon Hodson found that anti-vegetarian and anti-vegan bias is "commonplace and largely accepted" (736). Vegetarians and vegans, they point out, were more negatively ranked "when their motivations concern social justice rather than personal health" (732). The reasons for such bias, MacInnis and Hodson posit, include that vegetarians and vegans "undermine the integrity of prevailing social values and traditions that exploit animals" (739). They see such bias as similar to the bias against environmentalists and feminists who, similar to vegetarians and vegans, "do little objective harm but threaten the status quo." (739). Going against social norms that include using natural resources to raise animals for consumption, beauty products, and clothing, vegan and vegetarians become a threat to meat-eaters who want to continue "life as it is," "life as it should be," and "life as it's always been."

MacInnis and Hodson's (2017) discussion of vegan and/or vegetarian food choices is intricately connected to broader discussions on how we position ourselves within and outside specific discourse communities, how we see individual choices, and how we participate in structural frameworks that confirm, discourage, or openly discredit choices we make. Laura Hahn and Michael Bruner (2012), for example, point out in their discussion on organic, vegetarian, and vegan discourse conventions and politics that our consumer choices function as a vehicle to show "one's politics on one's plate" (42). Similarly, Chelsea Chuck, Samantha Fernandes, and Lauri Hyers (2016) argue that "politicized eaters engage in dietary practices that differ from the dominant food culture" (425). They point out that such politicized eaters include a wide range of participants, including labor organizers, animal rights activists, and international organizations such as *Stop Hunger Now*. These politicized members "participate in countercultural political acts through one of the most pervasive and intimate expressions of their culture and community – their food" (425). In addition, Abigail Seiler (2012), in her work on liberalism and Michelle Obama's obesity rhetoric, brings in

DOI: 10.4324/9781003039013-2

the problematic concept of individual choice when systemic inequalities are the underlying reason for eating habits and food choices. These discussions show the importance of contextualizing how we approach food choices, and what implications these choices have on individuals and systems. They also show that the multitudinous and sometimes contradictory rhetorical strategies surrounding discussions of food choices go beyond politics and encompass an understanding of how our positionalities influence and are influenced by normative behaviors entrenched in our political systems and social structures.

In this chapter, I reflect on the current divisions and controversies surrounding plant-based eating, veganism, and vegetarianism, and the role of ecofeminism in connecting human realities and ecological realities. I highlight the need for multiple and sometimes contradictory paradigms so that we can emphasize the complex, multidirectional, and trans-situational nature of the rhetorical choices that surround ecofeminism, veganism, vegetarianism, and plant-based food choices. I focus specifically on how our ideological, political, and social frameworks are influenced by, conflict with, and can expand normative constructs that define acceptable and unacceptable social and ecological behaviors. I contextualize and situate the conversation by first establishing the need for situational and ecological rhetorics to emphasize contextualized discussion from multiple and diverse theoretical and practical frameworks. I connect these discussions to the development of feminist thought and ecofeminist paradigms to show the fluidity of knowledge creation and the importance of trans-situational, transdisciplinary, and transnational alliances in our discussions of plant-based eating, vegetarianism, and veganism. I interrupt my theoretical discussions and connect narratives that highlight my own education as an ecofeminist and plant-based eater. With this, I show how ideologies connect to food choices, and how these choices are shaped by a complex mix of factors that I describe as a trans-situational and transnational counter-normative lifestyle where decisions are never easy and opportunities for re-evaluating my lifestyle choices are constant.

I consciously chose the concept of narrative interruption and continuation to show the interconnectedness of and also the struggles between theory and practice, individual action and systemic social transformation, where new facts, new theories, and new paradigms are constantly shaped and reshaped. With this, I follow Jean Francois Lyotard's (1984) postmodern argument that universal or "master narratives" need to be complemented and replaced by local narratives to include the voices and values of people who have been marginalized by normative and exclusive narratives reported by academics and non-academics. Here, I want to leave decontextualized rhetorical explorations and emphasize the situational, contextual, and ecological power of the rhetorical choices we make. I focus attention on the stage—the setting, the location, the scene—of the rhetorical event, to show new ways of discussing the importance of drawing connections when we read, interpret, create, refine, and apply diverse rhetorical lenses to our understanding of ecofeminism, veganism, vegetarianism, and plant-based eating.

My values and attitudes, and my efforts to reconsider current paradigms started when I listened—when I engaged in what Krista Ratcliffe (1999) called "interpretive invention"—to narratives told in my family and community, and later when I connected these narratives to explorations in rhetorical theory. I learned that my values are intricately connected to my complex cultural and social experiences as a farm-raised academic, woman, non-U.S. citizen, and a socially conscious individual invested in ecofeminism and plant-based eating. In other words, I realized the importance of expanding on and contributing to a transformation of universal narratives and normative writing conventions, and of joining together different strands of meaning-making to add new voices to the conversation. Through rhetorical listening (see Ratcliffe 1999, 2006), I also gained a new understanding of the need to interweave the personal with the theoretical to highlight the complexities and intersectionalities of current conversations on women, the environment, sustainability, and food choices. As Nicole Walker (2017, 10) puts it, "as you stitch an essay together, you stitch yourself into the world. The world, stitched by you, is made more whole."

How We Become: Emerging, Diverging, Converging

"Why would it be a problem to eat eggs?" my father wants to know when we talk about veganism in the big farm kitchen where vegans, vegetarians, plant-based eaters, and meat-eaters were gathering for lunch. "The family has raised chickens for more than a century, before and during the first World War, during the depression, during the second World War, and when we were raising six kids who needed to eat." He went on to ask: "Do you think it's better to eat some fruit that you can't find anywhere in Austria and that's shipped from who-knows-where? I know each chicken; I know when they lay their eggs; I know each rooster; I look for them at night if they haven't come up from the brook; they eat the bugs and worms and keep our soil healthy." We move to discussions of where our food comes from, and how farming is no longer done by small family farms but by corporations. "That's different," he says. "We raise what we eat. We eat what we raise. And we raise healthy and happy chickens who have the run of the outdoors. No chemicals. No fertilizers. No supplements." My father didn't label himself an organic farmer; he didn't call himself an environmentalist although his words and actions fit these terms; in fact, he denied that he was one because the words had little meaning to him. He collected rainwater in an old barrel. He built a windmill because he was interested to find out how much energy it would produce. He built a watermill for the same reason. And when solar panels became affordable, he convinced my brother to add solar energy to lower his dependence on the electric grid. "This is how you live. If you can give back, you give back," he would say. "I don't need a term for that."

Listening to my father's stories and learning about his insistence on treating the environment with care and respect instilled the same principles in his kids. It also instilled respect for my father whose knowledge about what can be grown in loamy soil, what needed dry earth, how to graft a fruit tree, or when to harvest all that was planted far surpassed my own knowledge of a healthy environment and healthy food choices. However, respect for the Earth, and respect for those who attempt to protect the Earth by choosing to be plant-based eaters, vegetarians, or vegans, is not as prevalent as we might hope. Two years after MacInnis and Hodson's (2017) publication of their study on bias against vegans and vegetarians was published, George Reynolds (2019), writing for *The Guardian*, asks, "Why do people hate vegans?" especially when "a consensus is starting to form that eating less meat would almost certainly be better for everyone – and the Earth." Reynolds agrees with many writers interested in food and animal studies (Adams 1994/2018; Bediako 2015; Bogueva and Marinova 2019; Bogueva et al. 2019; Frye and Brunner 2012; Gaard 2002; Gambert and Linné 2018; Hahn and Bruner 2012; Haraway 2003; Harper 2010, 2012; Quinn and Westwood 2018; Seiler 2012; Wright 2015, 2017) when he states that "what we grow, harvest, fatten and kill is political." He concludes that "the vegan wars are not really about veganism at all, but about how individual freedom is coming into conflict with a personal and environmental health crisis."

Reynolds' (2019) piece hints at the complexities that surround discussions of veganism, vegetarianism, and plant-based eating. Ecofeminists, animal studies theorists, and environmentalists, among others, have tried to uncover the connections between food choice and membership in specific social movements. Carol Adams (2018 "About"), for example, brings up "animal ecofeminism" which "challenged the idea that you could talk about the environment and consider yourself an environmentalist without addressing the fact that people were eating animals and dairy products and eggs." Similarly, vegetarian ecofeminism addresses speciesism, the oppression of nonhuman animals, and the importance of addressing the "political context of dietary choices" (Gaard 2002, 117). And Lori Gruen (2015) argues for rethinking our relationship with animals by advocating for an entangled empathy to strengthen our ethical agency.

These and other theoretical frameworks (see Birke and Hockenhull 2012; Donovan 1993, 2006; Warren 1987) are important building blocks for explaining how we understand the connections between environmental consciousness and food choices. With these general frameworks in mind, the next step is to participate in localized conversations that incorporate situational knowledge and with it enrich and transform our understanding of the world around us. Engaging in contextualized discussions about ecofeminism, veganism, vegetarianism, and plant-based eating, and situating them in lived experiences, ensures that we include instead of exclude, and that we take into consideration diverse experiences based on gender, culture, nationality, race, religion, politics, or age. Such a transformative localized approach will shift current discussions to include the experiences of individuals and groups currently at the periphery of ecofeminism, veganism, vegetarianism, and plant-based eating. A. Breeze Harper (2012), for example, in her discussions on the absence of race and class consciousness in the vegan movement, urges the vegan movement to "engage in dialogues that make visible the intersections of race-consciousness, socio-economic class, problematics of normative whiteness, and First World and class privileged perspective within American veganism" (157). Similarly, Michelle Loyd-Paige (2010) in an article published in *Sistah Vegan: Black Female Vegans Speak on Food, Identity, and Society*, explains that "in the U.S., how we treat food animals is reminiscent of how people of color were treated" (4) during slavery, and how people who move outside accepted norms are often treated today. Social media sites such as *Blacks Going Vegan* (2022) and *Black Vegans Rock* (2022), and a September 2019 Gallup survey reporting that African Americans eat 31 percent less meat (McCarthy and DeKoster 2020), show the need for more situated and contextualized research, where the scene of action—the values and attitudes we bring to the table—are included in our discussions and thus provide an opening to localize the theoretical and to theorize the local.

Situational rhetoric, made accessible through Lloyd Bitzer's 1968 article "The Rhetorical Situation," incorporates attention to the context in which rhetoric takes place. With its identifying and often seen as stratifying features of exigence, audience, and constraints, Bitzer's theory has received some criticism for excluding multiple rhetorical situations existing at the same time. Nevertheless, he brought attention to the importance of seeing rhetoric as situational, influenced by "an imperfection marked by urgency," an audience that can participate as change agents, and the limitations as well as the means available in a particular situation to shape an argument. Using this basic understanding of situational rhetoric, Craig Smith (2017) in his discussion of Bitzer's rhetorical theory, expands on the principles discussed by Bitzer and shows the importance of paying attention to the "multiplicity of the rhetorical situation" with many exigencies, diverse audiences and many constraints. Similarly, Jenny Edbauer (2005) reconceptualizes rhetorical situation or rhetorical publicness "as a context of interaction," adding "the dimensions of history and movement (back) into our visions/versions of rhetoric's public situations" (9). With this, she follows Michael Warner's (2002) argument of the public as created over time, through multiple texts and multiple voices, and Louise Weatherbee Phelps' (1988) discussion of rhetorical situation as part of historical and lived processes.

Edbauer's (2005) model of situational rhetorics, which she terms "rhetorical ecologies," encourages as to use the contextual concept of rhetorical situation and "add the dimension of movement back into our discussions of rhetoric" by paying attention to the wider ecology of rhetoric. This "ecological augmentation" (20) encourages us to pay attention to "trans-situationality" (20), a reference to the shifting and often conflicting theoretical and practical perspectives included in any rhetorical situation. As Nathaniel Rivers and Ryan Weber (2011) put it, looking at complex systems as ecological engages us to "contest, pull, push, and grab at one another" and relate to one another as social beings who are part of a system that is in constant flux (192). The expanded definition of rhetoric as situational, ecological, and ideological is especially important for augmenting our discussions of environmental movements, ecofeminist movements, and vegan and vegetarian movements where disagreements about who can claim membership, who can have a voice, and who can and should be excluded instead of included can be detrimental in our combined fight in the climate change crisis. Situational and ecological rhetorics is a much-needed approach if we want to include a multiplicity of voices and diverse lived experiences in our re-thinking of past and current theories and past and current narratives surrounding discussions of social change and how our food choices can encourage such change and impact normative hierarchical power structures.

The Practical Education of a Feminist, Ecofeminist, and Plant-Based Eater

Nature, when I was growing up in the small working-class farming community in Austria, was much revered, enjoyed, and also feared. We understood that we couldn't dominate nature. We understood that choices we made would be trumped by the choices that nature made. This meant that Carol Adams's (1994/2018) discussion of patriarchal dualisms, her call to end the unjustified domination of women and nature in *Neither Man Nor Beast*, and her argument on the social injustices connected to women and the environment required little proof for me. Even though my upbringing did not include words such as feminism, ecofeminist theory, environmentalism, or sustainability, I lived in the middle of all these words.

"Be ready in 10 minutes!" my mother would shout through the open window. We had finished *Mittagessen* (lunch, the main meal of the day), and I had just finished putting the dishes away in the Austrian farm kitchen. It was a cool fall day, and time to harvest the potatoes that would last us through the winter. Everybody was needed to make the back-breaking work less taxing. Three family members—the responsible mature ones who knew how not to hack potatoes in half by accident—were ready with their hoes. Since I was one of the three younger kids, my task involved picking the hoed potatoes and putting them into a bucket. Barefoot, I could wiggle my toes in the soil, push some recalcitrant potatoes close to my fingers, and even pick up small potatoes and drop them into the bucket with my toes curled around them. As long as I kept up my part of the task and didn't put my toes too close to the hoe, I could play and work at the same time.

The process had started the year before. Where the potatoes were now growing had been planted with corn the previous year. Corn stalks had been mulched into the soil in the fall, manure was added to the mix. In the early spring, the soil was ploughed into neat rows. In the meantime, we were getting the left-over potatoes from the previous year ready for planting. They were wrinkly and sprouted, and ready to be cut. "Keep at least one eye for each cut. Two are better. It won't grow otherwise," my mother would say. Planting the potatoes was done by hand, making sure they were spread evenly and put deep enough to have room for future hilling. Weeding became a daily event, and once the potatoes started sprouting, picking off potato bugs and feeding them to the chickens was part of every child's duty. Too many days of rain meant that the potatoes might rot in the soggy ground; too many sunny days might dry out the soil and decrease tuber yields.

We didn't own a big farm. We didn't sell any of the produce that we were growing, or eggs that our chickens were laying, or pigs that were feasting mostly on leftovers. We needed what we grew for survival. As a blacksmith, my grandfather and later my father made just enough to provide essential necessities. Most food, one of those necessities, was grown and raised on the land to keep 4 adults and 6 kids from going hungry. Understanding the soil, feeding it, maintaining it, and replenishing it were parts of my early education. Growing what we would eat and producing a healthy crop yield was essential for feeding the family. We always knew where our food came from, when it was ready for harvest, how it could be canned, fermented, dried, smoked, stored, how long it would keep in the cellar, and what we would need to do to prepare it for our next meal.

I didn't call myself a feminist or an ecofeminist when I was growing up because I didn't know what the words meant. I also wasn't aware of vegetarians or vegans in the small Austrian community where I was born and raised. In the 1960s and 1970s, the men and women in this small rural town worked the fields, drove tractors, and fed the livestock. It wasn't a place where I listened to discussions of equal rights and responsibilities. Kids were tasked with potato planting, corn shucking, chicken feeding, and apple harvesting. I was one of those kids. This meant that I only learned about feminist theory in the late 1980s when I was enrolled in a U.S. university course on early women's rights advocates and suffragists. English philosopher Mary Wollstonecraft's 1792 publication of "A Vindication of the Rights of Woman" became an entrance point to learning about how women used established paradigms to start their arguments for their own rights, and how they shifted those paradigms to include women's rights. As Wollstonecraft points out, "my main argument is built on this simple principle, that if she be not prepared by education to become the companion of man, she will stop the progress of knowledge and virtue" (Introduction). Women, Wollstonecraft argued, had a right to freedom, to education, and to equality because these rights would lead to a stable home, educated children, and a successful marriage.

Wollstonecraft's (1792) argument fascinated me. She wanted to expand the existing paradigm of late 18th century Britain that portrayed and treated women as emotional beings who were incapable of rational thought. Wollstonecraft's rhetorical decision to expand the definition of home and incorporate social and public life, and to argue for women's right to freedom and education, brings to light her understanding of the situational constraints—the barriers that more radical arguments would surely encounter. Her arguments and the arguments of early feminists were adapted by early women's rights defenders and women's voting rights supporters, revised and often derided by 20th century feminists, expanded to include women of all colors and backgrounds, and reconstituted to highlight different political, social, and cultural feminist and ecofeminist agendas. In the 21st century, after continuous struggles and fights for gender equality, it might seem easy to discard Wollstonecraft's argument. However, the Equal Rights Amendment (ERA) is still only a proposed amendment to the U.S. Constitution, 100 years after the 19th Amendment was passed. Even though most of us might no longer subscribe to late 18th-century philosopher Immanuel Kant's (1798) argument on men's natural supremacy, and the inability of women to reason and to actively take on moral responsibility, we are still faced with a continued belief that women are more emotional, more passive citizens, and, because of that, not ready to serve as the president of the United States. For example, as Abby Corrington and Michelle Hebl (2018) show in an article in *Equality, Diversity and Inclusion: An International Journal,* Hillary Clinton's run for the presidency was harmed by both hostile and benevolent sexism directed at women, social expectations of women as communal, and a need for continuing the status quo. In other words, we are not as far removed from Kant and 19th-century philosopher Georg Hegel as we would hope. Cited in Carol Adams' (1990/2015) *The Sexual Politics of Meat*, and in Susan Bordo's (1993) *Unbearable Weight*, Hegel (1821) reinforces the differences between men and women by pointing out that

> The difference between men and women is like that between animals and plants. Men correspond to animals, while women correspond to plants because their development is more placid and the principle that underlines it is the rather vague unity of feeling. When women hold the helm of government, the state is at once in jeopardy, because women regulate their actions not by the demands of universality but by arbitrary inclinations and opinions. Women are educated – who knows how? – as it were by breathing in ideas, by living rather than acquiring knowledge. The status of manhood, on the other hand, is attained only by the stress of thought and much technical exertion. (par. 166)

Hegel (1821), who in part of his writings argued for a concept of logic where thought and being were closely connected, emphasized the patriarchal dualism that I learned to eschew long before I embraced the general concept of ecofeminism.

Hegel's anthropomorphic sexism, and his comparison of women to plants, is a reminder that supporters of women's equality and feminist movements, current ecofeminists, and movements that challenge hegemonic structures had (and have) to fight an uphill battle with much opposition from the establishment. Members of 20th and 21st century cultural ecofeminism, for example, still use the concept of dualism to fight against politically, socially, racially, culturally, and gender-specific normative behaviors introduced and upheld by a white male hierarchy (see King 1981; Plumwood 2004; Ruether 1975; Spretnak 1982; Warren 2000). These dualisms, as Carol Adams (2018) reminds us in *Neither Man nor Beast: Feminism and the Defense of Animals*, often include "culture/nature; male/female; self/other; white/nonwhite; rationality/emotion" and "humans/animals" (71). According to Adams, ecofeminism in general terms encourages us to recognize that "the domination of the rest of nature is linked to the domination of women and that both dominations must be eradicated" (71–72). Since meat is a "symbol and celebration of male dominance," Adams (1990/2015, 13–14) argues for more egalitarian "plant-based economies" which, according to her research, allowed women to be largely self-sufficient and autonomous. The implied correlation between plant-based economies and women, and meat-based economies and men, is a good initial step in trying to understand the positionalities of women in current-day economies. However, it also establishes a chasm that leaves out the interwoven nature of economies and the complex cultural, gender, and social links that encourage conversations across bridgeable rifts.

Bridging Differences: Opportunities for Situational and Ecological Rhetoric

"I am vegan," said my niece from a small Austrian farming community where chickens, sheep, goats, and cows roam free before their inevitable end. She first moved to vegetarianism because, at the age of 18, she decided that she could no longer participate in eating what she walked past every day, photographed, fed, and petted. As she puts it, "the main reason for being vegetarian and now vegan has and will always be an ethical one. I don't want to be part of the torturing and the horror animals have to go through for no reason." She pointed out that her move from a mostly vegetarian to a vegan diet started when she saw a documentary on the fate of caged chickens. This put her off eggs and chickens forever. And, "having a less critical impact on our planet is a nice plus," she concludes. To reaffirm her commitment to the planet, she sent me a text that every vegan saves more than 100 animals a year, and that eating meat depletes our planetary resources much more rapidly than eating plants. When I asked her which five words she associates with being vegan, she didn't hesitate: "For me, it means compassion, plants, future, justice, life," she said. "How about you? Are you finally one of us?"

"I moved to a plant-based diet for the time being," I responded without committing to being one of them. Talking to my niece on social media an ocean away, I told her that fish is in my dietary plan if it's humanely sourced and approved by the Monterey Bay Aquarium's Seafood Watch. "I stopped eating red meat and chicken for ethical and also for environmental reasons, and dairy products for health reasons," I told her. Her question reminds me that we share some sensibilities but not all. "But how can you eat fish? They are living beings who get slaughtered just so that you can eat them," my niece wanted to know. "I know it. I'll be re-evaluating my food choices again soon," I tell her. I am a strong believer in the importance of changing normative behaviors that threaten the sustainability of our natural environment, and I agree that planetary resources are at their breaking point. "Most of our food has to be plant-based, locally sourced, organic, and humanely raised," I continue. It doesn't satisfy my niece.

My niece and I make our food choices because of ethical and environmental values. We do, however, differ in what this means to us. This provides a window into current theoretical debates on how we position ourselves within or outside an accepted or a newly emerging paradigm. My niece's statement that "I am vegan" shows her positioned at the center of veganism, a shift that she completed and that breaks with previously accepted norms and behaviors. It is a place of essential being. My statement that "I moved to a plant-based diet for the time being" makes it clear that I entered this space after occupying a different one, and that I am willing to leave other possibilities—in my case moving to veganism—open. I am not yet wholly convinced that the new system is right for me, and I assume that another system, adjusted for developing needs, might be part of my future. In other words, my niece and I are pursuing the same goals of conscientious eating and increased awareness of the resources we use. However, our reasons for, and even our approach to, talking about veganism and plant-based eating are part of different generational, cultural, and ideological paradigms.

In addition to reminding me of situational rhetoric and rhetorical ecology, our exchange brings to mind Thomas Kuhn's (1963) concept of paradigm shifts—how major changes to the ways of understanding come about. He especially focused on the common beliefs and assumptions that are part of a traditional conceptual model in the sciences. This model, or paradigm, functions because the majority of practitioners in the field agree that problems can best be solved by using common beliefs and assumptions. Once contradictions and inconsistencies arise, practitioners are forced to acknowledge that a shift to a new system is necessary to solve new problems. Kuhn points out that paradigm shifts are messy within specific fields, and that new conceptual models are disruptive, disorderly, and controversial. The need for a paradigm shift in our food habits, and in our interactions with the natural environment, in addition to being part of academic research in the sciences, humanities, and professional fields, has been shown in news outlets, on social media, and in political debates surrounding the 2020 presidential elections, and in a majority of the discussions at the Democratic National Convention (2020). It is through the lens of situational rhetorics and rhetorical ecologies, with a strong focus on an analysis of the contextual information in its historical framework, that we can localize and theorize the disorderly and controversial process of such a paradigm shift.

Ecofeminist, environmentalist, and vegan, vegetarian, and plant-based movements that engage in the disorderly and controversial process of expanding and reconstituting existing paradigms, and with it accepted normative behaviors about the environment and our food choices, have been embraced for different reasons by politicians, individuals from different cultural and ethnic backgrounds, members of different generations, climate change activists, non-profit organizations, and for-profit businesses. An overwhelming majority at the 2020 Democratic National Convention, and with them Bernie Sanders (2020), for example, are calling out "the enormous threat to our planet of climate change" (Sanders 2020); Greta Thunberg (2019) fights for the survival of the planet, started Fridays for Future, and went vegan; the *Earth Day Network* (2022), a movement to "drive transformative change for people and planet," tells us to take action and "fight climate change with diet change;" *Food & Trees for Africa* (2022) is working toward long-term sustainability by addressing food security, environmental sustainability, and greening efforts through education and through training students, teachers and community members in Southern Africa; Native American tribes and environmentalists fight for environmental justice (Kennedy 2020); Helen Mountford (2020), Vice President of *Climate and Economics at World Resources Institute* and Program Director of the *New Climate Economy*, proposed in an article titled "From COVID to Climate: This convergence of crises calls for a convergence of solutions" that "we must align stimulus packages with ambitious policies that tackle climate change, environmental damage, and social inequalities" (Mountford); Microsoft and Google increased their renewable energy portfolios; Walmart is collaborating with environmental groups and suppliers on Project Gigaton to "cut a billion tons of greenhouse gas pollution from the company's global supply chain by 2030" (Murray 2019); *Delhi Greens* (2022), co-founded by Govind Singh and Ravinder Bawa, is focused on empowering young people to take green action; and Green Worker Cooperatives (2022) serves immigrants and communities of color to "build, grow, and sustain worker-owned green businesses to create a strong, local, and democratic economy rooted in racial and gender equity." The trans-situationality, as well as transdisciplinarity and transnationality, that defines these disparate sets of individuals and organizations, with many different reasons for promoting climate change solutions, changes to our food choices, and a need for conscious living, shows the importance of a contextualized rhetorical approach that acknowledges Edbauer's (2005) call for rhetorical ecologies where exigencies shift, individuals and groups can be influenced to change their allegiances, and situational factors that affect the expediency of the message can be turned on its side.

A situational and ecological rhetorical lens is especially appropriate when we explore women's role in society, the role of the natural environment and resource depletion, and the impact of food choices on the planet. Because of shifting social, political, and cultural needs, these movements have also shifted over the years, including a number of feminist "waves," ecofeminist lenses, environmental approaches, and food movements. Unfortunately, what we can overwhelmingly see in recent discussions on these topics is a propensity to dismiss historical and situated information, instead focusing on what was left out, what was ignored, and what was plain wrong. In other words, many of the discussions do not allow for a "fluid framework of exchanges" (Edbauer 2005, 19) where knowledge is created and shared by diverse participants with diverse purposes and values through a complex set of situational and localized events. Acknowledging the fluidity of knowledge creation, and the ability of participants to move into and out of diverse groups and movements, ensures that social justice groups collaborate, and that trans-situational, transdisciplinary, and transnational alliances are not only acknowledged but actively sought.

The Re-Education of an Ecofeminist: Theory, Application, and Practice

Many years after I had left the farm for work and life in the United States, I brought my newly acquired theoretical knowledge of food politics, environmentalism, and feminism back to the farm. It didn't go so well. "I read this book on feminism and the politics that go into meat production," I say. "Really? What did the book tell you?" my brother who had inherited the farm wanted to know. I tell him about Carol Adams' argument that if we wanted to halt the environmental downward spiral, we needed to rethink how we treated the natural world, how we treated women, and how we treated animals. I bring up the links that Adams saw between meat eating and masculinity. And I also talk about her explanation that women, who she calls "feminist-vegetarians," became vegetarians because they saw it as an active choice that went against the social and political norm of eating meat.

"Is that right? You read about this?" says my brother, watching a baby-calf that he helped birth the night before. "She is pretty weak. I was up all night with them. The mother won't let her drink yet. I need to milk her soon to get some nourishment into the little one." I look at the little calf and go on, not really noticing the irony of bringing up this topic at the family farm where I helped raise chickens, feed cows, grow potatoes, and harvest cherries, peaches, plums, and apples. "It's all connected. The environmental effect of raising animals is huge. And then they get slaughtered and we never see the actual animal when we buy pieces of meat at the market." I took my cue from Adams' discussion of the interrelationship among maintenance (the environmental effects of raising animals), production (the practice of raising animals for slaughter), and consumption (the act of buying and eating a hamburger) (78–83). Adams argues that we need to understand the consequences of meat consumption so that we learn about "the loss of topsoil, water, and the demands on fossil fuels that corpse production requires" (p. 78). "And most people don't even know where their food is coming from. That's why it's so easy for them to eat meat." I said forcefully.

My brother turned and leaned against the railing. "I know where my food is coming from," he said. "Of course it's all connected. You can't grow up on a farm and think otherwise. See the little one?" He turned around again to check on the baby-calf. "All this," he pointed to the two cows and the calf grazing in the meadow, to the chickens, to the vegetable plots, the newly seeded trees that would replace the trees near the Danube that didn't survive the flood, "is part of the ecosystem. You know that. You were part of this ecosystem. Everything has a place. I take care of the soil; I plant the potatoes; I cut the grass to have hay in the winter for the cows; I check on the woods to make sure we don't have bark beetles. I pay attention to the weather; I compost everything that can be composted. Your father built the windmill from the bottom up. And 20 years later he added solar panels to the roof. I collect rain water so that we don't run out if the well dries up. If I don't take care of the soil and replenish it, and if I overplant, I won't have enough food to take us through the winter. The chickens are helping with bug control and are laying some beautiful eggs. A few just hatched. Do you want to see them?" he asked. "What about the cows?" I want to know as we walk to the chicken coop. "Wouldn't it be better to sell the cows and use the meadow for vegetables?" He picks up one of the 25 new chicks and hands it to me. "I count them every night to make sure they are all in the coop. We have foxes that will get them otherwise." The mother hen is not happy with me and I put the chicken back on the ground. "Well, I have thought about it," he continues. "Right now, I have the land to graze them. I am not doing it for profit. They are 100% grass fed. And I have thought of a few goats or sheep in the near future. They are easier to handle." He looks around. "It's not a question of vegetables or meat. It's a question of what you do with it, how you treat the land, the animals, what life you give them. Sure, I don't need to raise animals, or eat meat. I hardly eat meat any longer, actually." We walk back to check on the baby-calf. "Remember when our parents bought two baby sheep, and you named them?" I remember very well. "And how mad our mom was when you cried your eyes out when the sheep were no longer grazing in the fields?" It is very much still part of the narrative I use to talk about my unwillingness to eat mutton stew at Native festivals. "Well, we haven't named the animals since," he tells me.

"With my two or three cows, I can let them roam in the meadow. I don't need to cut down any forest," he continues. "Austria has strict regulations about cutting trees. Every tree you cut needs to be replanted." We talk about planting hundreds of tiny trees our father grew from seed on a steep hill near the house more than 30 years ago, not because we had to, but because my father saw it as a no-brainer to plant something that would halt the erosion of soil and that could provide shade for the chickens and geese roaming around and feeding on snails and worms and beetles. "The water for them is rainwater. They have a healthy life, and it's pretty low-impact on the environment. Sure, I know about methane. It's not easy to make decisions about what's best. I talked to my neighbor, and their kids live in the city. They don't care for anything homegrown. They'd rather go to the store and buy food that's already made from who knows where. Carbon dioxide is a problem too, and the huge corporate agriculture isn't helping. Maybe that's what your book was talking about. If that's what she knows, that's what she knows. Here is what I know now, and who knows what I'll know in the future," he said, helping the baby-calf stand up and take its first steps.

My brother's perspectives, his knowledge of the land, and his concern for the environment brought to the forefront the dangers of a single story (Adichie 2009), and the dangers of imposing my situated ideological framework on others. Instead, the power of trans-situational exchanges, of exchanges that promote understandings across disciplines and nations and across easily established divides, can encourage conversations that focus on learning from each other, on paying attention to common ground without erasing differences, and on creating new alliances that can help us counteract the many challenges we currently face, and that can also help us see the connections among environmental degradation, food choices, climate crisis, racial divides, and gender and LGBTQIA discrimination.

Calling myself an ecofeminist signifies a political stance, a counter-hegemonic ideology, and a way to identify myself with and against other social, political, environmental, cultural, and feminist movements. What is problematic for me is that I am often expected to define myself against what is seen as a normative patriarchal system, my upbringing on a farm, and my father's and brother's attempts to feed the family, pay attention to the land, and create spaces that accommodated all of us, with animals, vegetables, legumes, nuts, and fruits harvested from the land. In addition, I am also encouraged to take a stance against certain branches of feminists, environmentalists, ecofeminists, vegans, and vegetarians who approach women's rights for equal humanity and animal rights from different theoretical, philosophical, and rhetorical perspectives. For example, as critics of ecofeminism point out, there are too many variations of ecofeminisms, with some seemingly irreconcilable differences. Ecofeminisms grounded in philosophy, in politics, and in ethics, including cultural ecofeminism, social ecofeminism, socialist ecofeminism, vegetarian ecofeminism, exemplify the potential for schisms and the need for plural ecofeminisms. In other words, my cultural background, my country of origin, my politics, my ethical perspectives on nonhuman nature, my religious beliefs, and my food choices could alienate ecofeminists whose backgrounds and perspectives are different from mine.

Elizabeth Carlassare (1994), in her discussion on essentialism and ecofeminism, shows, for example, the divisions between social/ist ecofeminists such as Janet Biehl, Alison Jaggar, Audre Lorde, and Donna Haraway, and cultural ecofeminists such as Riane Eisler (1987), Merlin Stone (1976), Gloria Orenstein, (1990), Susan Griffin (1978), and Mary Daly (1978). Social/socialist ecofeminists argue that cultural ecofeminists employ essentialist notions of women instead of emphasizing the construction of gender, race, and sexuality. As Carlassare (1994) notes in support of social/socialist ecofeminists, essentialized notions of gender "allows women with race, class, or national privileges to sidestep their obligation to take responsibility for their own power and participation in structures of domination" (229).

Despite the potential conflicts in ecofeminist positions, Carlassare (2000) argues that the absence of a unified philosophical or ideological position actually creates an opportunity for a social movement that focuses on gender equality, social change, and planetary survival. As Carlassare puts it, "most ecofeminists do not strive to arrive at a unified, coherent, epistemological position; there is no one philosophy or theory that defines ecofeminism, rather there are differently situated, overlapping, and interacting outlooks that have emerged under ecofeminism's umbrella" (100). This promotes "multiplicity of perspectives" (103) and encourages participants to "unite and work together for social change while making sufficient space for different points of view" (102). It also takes into account the diverse political, cultural, social, and ecological contexts in which ecofeminist activism takes place, leaving behind essentialist notions such as gender, race, and sexuality. According to Carlassare, paying attention to "context dependency and diversity" enables ecofeminists to "become a global network of resistance against ecological destruction and the legacies of patriarchy, capitalism, and imperialism" (102).

My niece sees saving the planet and veganism as closely connected, and she tells me that she sees herself as a member of a group "that speaks for the animals, because they have the right to live like everyone else." She continues to point out that "cows and chickens are women and mothers too." And she continues, "Being vegan reduces the negative impact on our planet a lot. The livestock industry needs a ridiculous amount of water, food and other resources. Skipping that step and eating the plants instead of feeding it to animals that are then eaten is better for our planet should be common sense." My niece's convictions closely align with Carol Adams's (1990/2015) understanding of ecofeminism and veganism. This position, however, pushes aside my father's and my brother's understanding of how the human and nonhuman world interact with each other. With this in mind, I am no longer focused on fighting the Cartesian dualisms outlined by cultural ecofeminists; instead, I agree with Elizabeth Carlassare's (1994) and Val Plumwood's (2004) discussions of critical eco-feminism. Plumwood makes sure to point out that "women are no more 'part of nature' or 'closer to nature' than men are – both men and women reside in both nature and culture. We must rethink hyper-separations, both the opposition and polarisation of men and women and that of humanity and nature" (50). She continues her argument by pointing out that "eco-feminism promotes ecological consciousness which insists that truly human life is embedded in both nature and culture, which are not hyper-separated spheres as Artemisian feminism assumes" (51). Providing opportunities for inclusion instead of focusing on reasons for separation increases our chances of changing established societal norms and encouraging collaborations among diverse stakeholders. For ecofeminists, then, instead of feeling beleaguered by competing branches of feminism and ecofeminism, this means that we can focus on addressing situated practices related to environmental degradation, food security, sustainable farming practices, women's diverse roles in localized communities and global efforts to address equal rights. It allows us to focus our agency and ambition not by joining forces but by diversifying forces.

In 1993, Greta Gaard already emphasized in "Living Interconnections with Animals and Nature" that ecofeminism is an interdisciplinary initiative that draws from such fields as ecology, feminism, and socialism. According to Gaard, ecofeminism's premise is that the ideology which authorizes oppressions such as those based on race, class, gender, sexuality, physical abilities, and species is the same ideology which sanctions the oppressions of nature. To address the global challenges we are facing, she encourages us in *Critical Ecofeminism* (2017) that the collaboration among "the movements for environmental justice, ecofeminism, and sustainability" is of utmost importance (4). This proposal necessitates an understanding of and respect for the many contexts, values, and situational practices that members of these movements bring to the table. Foregrounding commonalities by highlighting intersectional approaches and inclusive practices can lead to a contextualized examination of the many normative cultural and social behaviors that, in the Western world, are based on a largely patriarchal, heterosexual, white, male, middle-class system. Including different cultural beliefs is especially important and, according to Gaard (2010), needs to be examined by ecofeminisms to address "the meaning and practice of activism, and the rhetorical strategies best suited for developing and communicating feminist and ecofeminist perspectives in specific cultural context." (11). This encourages us to learn from each other, to listen, and to make sure that we appreciate diverse cultural and social ideologies. As Gaard (2017) puts it, critical ecofeminism encourages "the practice of attentive listening" (xvi). This, she points out, is part of a broader feminist methodology which "emphasizes *listening* as a hallmark of good scholarship – listening to one's research subjects, to the oppressed, to one's activist and scholarly community" (xvii).

The kind of listening suggested by Gaard (2017) is, with an accompanying productive silence, according to Glenn and Ratcliffe (2011), "essential preparation for speaking" (2). As Glenn and Ratcliffe (2011) convincingly argue in their introduction to the collection on *Silence and Listening as*

Rhetorical Arts, individuals and groups "can more productively discern and implement actions that are more ethical, efficient, and appropriate when all parties agree to engage in rhetorical situations that include not only respectful speaking, reading, and writing but also productive silence and rhetorical listening, all of which help prepare a person for eloquence" (3). In earlier publications, Ratcliffe defined rhetorical listening as "a trope for interpretive invention" (1999, 195) that "signifies a stance of openness that a person may choose to assume in relation to any person, text, or culture" (2006, 17). When engaging with difference, such listening encourages us to "generate more productive discourses about and across both commonalities and differences" (1999, 220). It also allows us, as Gaard (2011) indicates in her discussion of listening to early feminists and ecofeminists, to learn from a diverse body of knowledge and adjust it to highlight the intricate connections of the environment to multiple and intersectional forms of oppression and social injustices.

The Function of Expansion: A Proposal for Trans-Situational, Transdisciplinary, and Transnational Conversations

Critics of cultural ecofeminism have pointed out that it leaves out much of the situational context that surrounds discussions of food choices, ecofeminism, and environmentalism. The de-contextualized dualisms addressed in Adams, in other words, don't show the situational complexity of human interactions with each other and with the natural environment. This becomes especially problematic when our fight against dualisms includes dualistic solutions (e.g., vegetarian/meat-eater) which go largely unexamined. However, instead of devaluing Adams' and other's contributions, we can listen to the varied approaches of early proponents of ecofeminism and learn how to situate ourselves within and outside specific frameworks of thought, and how to use our own situated perspectives to create and establish intersectional arguments. It will encourage us to escape what Rivers and Weber (2011), in their work on ecological pedagogy and rhetoric, see as our propensity to work within the known. As they point out, we are often "in favor of acknowledging and responding within frameworks of existing norms, regulations, and procedures, even when an advocate's ultimate goal is to drastically change those frameworks" (190). Changing an existing structure or paradigm, in other words, is never easy and requires close attention to how and why discriminatory and oppressive structures and paradigms came into existence and became established institutional strongholds that are difficult to escape.

Understanding our inclination to remain part of an existing system—to avoid messy, disruptive, and controversial paradigm shifts (Kuhn 1963)—and being aware that our positionalities and ideologies are often influenced by a hegemonic, hierarchical, and patriarchal structure is necessary to grasp the difficulties of changing the frameworks that have defined us for much of our lives. Veganism and vegetarianism, for example, are often defined through current paradigms that are guided by a normative meat-eating majority and that show what it is not. *The Vegan Society* (2022) sees veganism as "a way of living which seeks to exclude, as far as is possible and practicable, all forms of exploitation of, and cruelty to, animals for food, clothing or any other purpose." Similarly, *The Vegetarian Society* (2022) tells us that "Vegetarians don't eat fish, meat or chicken. Vegetarians and vegans don't eat products or by-products of slaughter. They don't eat any foods which have been made using processing aids from slaughter." Furthermore, we are enticed to purchase "meat substitutes," including vegetarian burgers, vegetarian bacon, vegetarian hotdogs, and tofurky. To shift the discussions from *what is not* to *what is* can be achieved by acknowledging an expanded definition of what falls within and outside normative behaviors. Focusing on what I *do* eat (organic, homegrown, sustainably farmed vegetables, fruits, legumes, nuts, grains), instead of focusing on what I *don't* eat (meat) pays attention to a new stage—a stage that shifts attention and power from a narrowly defined normative to a new normative where vegan, vegetarians, and plant-based eaters are part of trans-situational rhetorical events that acknowledge the contextual choices that we make.

Expanding how we situate ourselves within and outside movements, who we see as members of our communities, and how we interact with multiple groups should mean that we can share what we have in common, accept what we can't agree on, and listen and learn about new ways of seeing connections among food choices, environmental degradation, sustainability, food deserts, family farming, and even corporate farms. When we learn to appreciate multiple frameworks we can encourage the co-existence of multiple positions related to food choices or to ecofeminism; more importantly, we can expand and complicate our own positions, acknowledging that we are social beings and part of "complex systems, through which people relate and … contest, pull, push, and grab at one another" (Rivers and Weber 2011, 192). Carlassare (2000) applies the principles of expanding perspectives and integrating multiple positions to ecofeminism by emphasizing the need to bridge divides among cultural ecofeminists and social/ist ecofeminists instead of emphasizing those divides. In addition to heeding Carlassare's call for embracing multiple ecofeminist perspectives, we can add the importance of seeing these perspectives within the historical development of social movements and their intricate connections to shifting political, racial, cultural, and environmental concerns. The long history of feminism and ecofeminism, and the interconnections between ecofeminism and veganism, vegetarianism, and plant-based eating, for example, show how rhetorical ecologies allow us to adapt, expand, and transform situated knowledge that is in continuous "social flux" (Edbauer 8) In other words, we need to make sure that we don't leave knowledge behind or dismiss it or push it aside, but that we push knowledge forward, revisit it, revise it, integrate it, and expand it.

Bibliography

Adams, Carol J. 1994/2018. *Neither Man nor Beast: Feminism and the Defense of Animals*. New York, NY: Bloomsbury.

Adams, Carol J. 1990/2015. *The Sexual Politics of Meat: A Feminist-Vegetarian Critical Theory*. 25th Anniversary Edition. New York, NY: Bloomsbury.

Adams, Carol J. 2018. "About Ecofeminism." Accessed August 27, 2020. https://caroljadams.com/about-ecofeminism

Adichie, Chimamanda N. 2009. *The Danger of a Single Story. Ted Talk*. https://www.ted.com/talks/chimamanda_ngozi_adichie_the_danger_of_a_single_story

Bediako, Jacqueline. 2015. Food Apartheid: The Silent Killer in the Black Community. *The Atlanta Black Star*, June 16, 2015. http://atlantablackstar.com/2015/06/16/food-apartheid-the-silent-killer-in-the-black-community/

Biehl, Janet, 1991. *Rethinking Ecofeminist Politics*. Boston, MA: South End Press.

Birke, Lynda, and Jo Hockenhull, eds. 2012. *Crossing Boundaries: Investigating Human-Animal Relationships*. Leiden, Netherlands: Koninklijke Brill.

Bitzer, Lloyd F. 1968. "The Rhetorical Situation." *Philosophy & Rhetoric* 1, no. 1: 1–14.

Black Vegans Rock. Accessed March 1, 2022. https://www.blackvegansrock.com/#home-section

Blacks Going Vegan: Promoting Health Building Plant-Based Nutrition for African Americans. Accessed March 1, 2022. http://blacksgoingvegan.com/

Bogueva, Diana, and Dora Marinova. 2019. "Reconciling Not Eating Meat and Masculinity in the Marketing Discourse for New Food Alternatives." In *Environmental, Health, and Business Opportunities in the New Meat Alternatives Market*, edited by Diana Bogueva et al., 260–282. Hershey, PA: IGI Global.

Bogueva, Diana, Dora Marinova, Talia Raphaely, and Kurt Schmidinger, eds. 2019 *Environmental, Health, and Business Opportunities in the New Meat Alternatives Market*. Hershey, PA: IGI Global.

Bordo, Susan. 1993. *Unbearable Weight: Feminism, Western Culture and the Body*. Berkeley, CA: University of California Press.

Carlassare, Elizabeth. 1994. "Destabilizing the Criticism of Essentialism in Ecofeminist Discourse." *Capitalism Nature Socialism* 5, no. 3: 50–66.

Carlassare, Elizabeth. 2000. "Socialist and Cultural Ecofeminism: Allies in Resistance." *Ethics and the Environment* 5, no. 1: 89–106.

Chuck, Chelsea, Samantha A. Fernandes, and Lauri L. Hyers. 2016. Awakening to the Politics of Food: Politicized Diet as Social Identity. *Appetite* 107: 425–436.

Corrington, Abby, and Michelle Hebl. 2018. "America Clearly Is Not Ready for a Female President: Why?" *Equality, Diversity and Inclusion: An International Journal* 37, no. 1: 31–43.

Daly, Mary. 1978. *Gyn/Ecology: The Meta-Ethics of Radical Feminism*. Boston, MA: Beacon Press.

Delhi Greens. 2022. "About Us." Accessed March 1, 2022. https://delhigreens.org/about/

DeLuca, K. M. 1999. *Image Politics: The New Rhetoric of Environmental Activism*. New York, NY: Routledge.

Donovan, Josephine. 1993 "Animal Rights and Feminist Theory." In *Ecofeminism: Women, Animals, Nature*, edited by Greta Gaard, 167–194. Philadelphia, PA: Temple University Press.

Donovan, Josephine. 2006. "Feminism and the Treatment of Animals: From Care to Dialogue." *Signs: Journal of Women in Culture and Society* 31, no. 2: 305–329.

Earth Day Network. 2022. "Fight Climate Change with Diet Change." Accessed March 1, 2022. https://www.earthday.org/actions/fight-climate-change-with-diet-change/

Edbauer, Jenny. 2005. "Unframing Models of Public Distribution: From Rhetorical Situation to Rhetorical Ecologies." *Rhetoric Society Quarterly* 35, no. 4: 5–24.

Eisler, Riane Tennenhaus. 1987. *The Chalice and the Blade: Our History, Our Future*. New York, NY: HarperCollins.

Food and Trees for Africa. 2022. "About Us." Accessed March 1, 2022. https://trees.org.za/what-we-do/

Frye, Joshua, and Michael S. Brunner, eds. 2012. *The Rhetoric of Food: Discourse, Materiality, and Power*. New York, NY: Routledge.

Gaard, Greta. 1993. "Living Interconnections with Animals and Nature." In *Ecofeminism: Women, Animals, Nature*, edited by Greta Gaard, 1–12. Philadelphia, PA: Temple University Press.

Gaard, Greta. 2002. "Vegetarian Ecofeminism: A Review Essay." *Frontiers: A Journal of Women Studies* 23, no. 1: 117–146.

Gaard, Greta. 2010. "New Directions for Ecofeminism: Toward a More Feminist Ecocriticism." *ISLE: Interdisciplinary Studies in Literature and Environment* 17, no. 4: 643–665.

Gaard, Greta. 2011. "Ecofeminism Revisited: Rejecting Essentialism and Re-placing Species in a Material Feminist Environmentalism." *Feminist Formations* 23, no. 2: 26–53.

Gaard, Greta. 2017. *Critical Ecofeminism*. New York, NY: Lexington Books.

Gambert, Iselin and Tobias Linné. 2018. "From Rice Eaters to Soy Boys: Race, Gender, and Tropes of 'Plant Food Masculinity.'" *Animal Studies Journal* 7, no. 2: 129–179.

Glenn, Cheryl, and Krista Ratcliffe. 2011. "Introduction: Why Silence and Listening Are Important Rhetorical Arts." In *Silence and listening as rhetorical arts*, edited by Cheryl Glenn and Krista Ratcliffe, 1–19. Carbondale, IL: Southern Illinois University Press.

Green Worker Cooperatives (2022). "Mission." Accessed March 1, 2022. http://www.greenworker.coop/mission

Griffin, Susan. 1978. *Woman and Nature: The Roaring Inside Her*. New York, NY: Harper and Row.

Gruen, Lori. 2015. *Entangled Empathy: An Alternative Ethic for Our Relationships with Animals*. Brooklyn, NY: Lantern Books.

Hahn, Laura K. and Michael S. Bruner. 2012. "Politics on Your Plate: Building and Burning Bridges across Organic, Vegetarian, and Vegan Discourse." In *The Rhetoric of Food: Discourse, Materiality, Power*, edited by Joshua Frye and Michael S. Brunner, 42–57. New York, NY: Routledge.

Haraway, Donna. 1991. *Simians, Cyborgs, and Women: The Reinvention of Nature*. New York, NY: Routledge.

Haraway, Donna. 2003. *The Companion Species Manifesto: Dogs, People, and Significant Otherness*. Chicago, IL: Prickly Paradigm Press.

Harper, A. Breeze. 2010. *Sistah Vegan: Black Female Vegans Speak on Food, Identity, Health, and Society*. Brooklyn, NY: Lantern Books.

Harper, A. Breeze. 2012. "Going Beyond the Normative White 'Post-racial' Vegan Epistemology." In *Taking Food Public: Redefining Foodways in a Changing World*, edited by Psyche Williams-Forson and Carole Counihan, 155–174. New York, NY: Routledge.

Hegel, Georg W. F. 1821. *Philosophy of Rights*. Republished 1833, 1854. Translated by Alan White, 2002. Indianapolis, IN: Hackett.

Jaggar, Alison M. 1983. *Feminist Politics and Human Nature*. Totowa, NJ: Rowman and Allanheld.

Kant, Immanuel.1798. *Anthropology from a Pragmatic Point of View*. Translated and edited by Robert B. Louden, 2006. New York: Cambridge University Press.

Kennedy, Merrit. 2020. "Judge Orders Environmental Review of Controversial Dakota Access Pipeline." *NPR*, March 25, 2020. https://www.npr.org/2020/03/25/821643911/judge-orders-environmental-review-of-controversial-dakota-access-pipeline

King, Ynestra. 1981. "Feminism and Revolt." *Heresies* 4, no. 1: 12–26.

Kuhn, Thomas S. 1963. "The Essential Tension: Tradition and Innovation in Scientific Research." In *Scientific Creativity: Its Recognition and Development*, edited by Calvin Taylor and Frank Barron, 341–354. New York, NY: Wiley.

Lorde, Audre. 1984. *Sister Outsider*. Freedom, CA: Crossing Press.

Loyd-Paige, Michelle. 2010. "Thinking and Eating at the Same Time: Reflections of a Sistah Vegan." In *Sistah Vegan: Black female Vegans Speak on Food, Identity, Health and Society*, edited by A. Breeze Harper, 1–7. New York, NY: Lantern.

Lyotard, Jean-François. 1984. *The Postmodern Condition: A Report on Knowledge*. Minnesota, MN: University of Minnesota Press.

MacInnis, Cara C., and Gordon Hodson. 2017. "It Ain't Easy Eating Greens: Evidence of Bias Toward Vegetarians and Vegans from Both Source and Target." *Group Processes & Intergroup Relations* 20, no. 6: 721–744.

McCarthy, Justin, and Scott DeKoster. 2020. "Nearly One in Four in U.S. Have Cut Back on Eating Meat." Gallup, January 27, 2020. https://news.gallup.com/poll/282779/nearly-one-four-cut-back-eating-meat.aspx

Mountford, Helen. 2020. "From COVID to Climate: This Convergence of Crises Calls for a Convergence of Solutions." Organisation for Economic Co-operation and Development, June 5, 2020. https://www.oecd-forum.org/posts/from-covid-to-climate-this-convergence-of-crises-calls-for-a-convergence-of-solutions

Murray, Tom. 2019. "The Businesses that Are – And Are Not – Leading on Climate Change." *Forbes Magazine*, November 8, 2019. https://www.forbes.com/sites/edfenergyexchange/2019/11/08/the-businesses-that-are–and-are-not–leading-on-climate-change/#57a6438b7aa1

Orenstein, Gloria F. 1990. *Reflowering of the Goddess*. New York: Pergamon.

Phelps, Louise. W. 1988. *Composition as a Human Science*. New York, NY: Oxford University Press.

Plumwood, Val. 2004. "Gender, Eco-Feminism and the Environment." In *Controversies in Environmental Sociology*, edited by Rob White, 43–60. New York, NY: Cambridge University Press.

Quinn, Emelia, and Benjamin Westwood, eds. 2018. *Thinking Veganism in Literature and Culture: Towards a Vegan Theory*. Cham, Switzerland: Springer.

Ratcliffe, Krista. 1999. "Rhetorical Listening: A Trope for Interpretive Invention and a 'Code of Cross-Cultural Conduct.'" *CCC* 51, no. 2: 195–224.

Ratcliffe, Krista. 2006 *Rhetorical Listening: Identification, Gender, Whiteness*. Carbondale, IL: Southern Illinois University Press.

Reynolds, George. 2019. Why Do People Hate Vegans? *The Guardian,* October 25, 2019. https://www.theguardian.com/lifeandstyle/2019/oct/25/why-do-people-hate-vegans

Rivers, Nathaniel A., and Ryan P. Weber. 2011. "Ecological, Pedagogical, Public Rhetoric." *College Composition and Communication* 63, no. 2: 187–218.

Ruether, R. 1975. *New Woman, New Earth.* Minneapolis, MN: Seabury.

Sanders, Bernie. 2020. "Speech to the Democratic National Convention." Available from *NPR*, August 17, 2020. https://www.npr.org/2020/08/18/903398536/watch-bernie-sanders-speech-to-the-democratic-national-convention

Seiler, Abigail. 2012. "Let's Move: The Ideological Constraints of Liberalism on Michelle Obama's Obesity Rhetoric." In *The Rhetoric of Food: Discourse, Materiality, and Power*, edited by Joshua Frye and Michael S. Brunner, 168–183. New York, NY: Routledge.

Smith, Craig. 2017. *Rhetoric and Human Consciousness: A History.* Long Grove, IL: Waveland Press.

Spretnak, C. 1982. *The Politics of Women's Spirituality.* New York, NY: Doubleday.

Stone, Merlin. 1976. *When God Was A Woman.* New York, NY: Harcourt.

The Vegan Society. (2022). "Definition of Veganism." Accessed March 1, 2022. https://www.vegansociety.com/go-vegan/definition-veganism

The Vegetarian Society (2022). "What Is A Vegetarian?" Accessed March 1, 2022. https://vegsoc.org/info-hub/definition/

Thunberg, Greta. 2019. *No One Is Too Small to Make a Difference: Illustrated Edition.* New York: Penguin Books.

Walker, Nicole. 2017. "The Braided Essay as Social Justice Action." *Creative Nonfiction* 64: 6–10.

Warner, Michael. 2002. "Publics and Counterpublics." *Public Culture* 14, no. 1: 49–71.

Warren, Karen J. 1987. "Feminism and Ecology: Making Connections." *Environmental ethics* 9, no. 1: 3–20.

Warren, Karen. 2000. *Ecofeminist Philosophy: A Western Perspective on What It Is and Why It Matters.* Oxford, England: Rowman & Littlefield.

Wollstonecraft, Mary. 1792. "A Vindication of the Rights of Woman: With Strictures on Political and Moral Subjects." Gutenberg Collection, https://www.gutenberg.org/ebooks/3420

Wright, L. 2015. *The Vegan Studies Project: Food, Animals, and Gender in the Age of Terror.* Athens, GA: University of Georgia Press.

Wright, L. 2017. "Introducing Vegan Studies." *ISLE: Interdisciplinary Studies in Literature and Environment* 24, no. 4: 727–736.

3 Becoming Vegan

A Mixed-Methods Study of Vegan Identities

Silke Feltz

To date, many of the factors involved in self-identification of veganism or vegetarianism have focused on specific kinds of motivations such as ethical concerns, religion, socio-demographics, gender, or more general attitudes toward animals. While these factors are important, there is likely to be more than one way that could result in becoming vegetarian or vegan. And, within these different paths, there are likely to be multiple, intersecting factors that can be better understood rhetorically through a narrative lens in particular. This chapter proposes to explore some of these potential paths and some of the potential intersecting factors that lead to people self-identifying as vegetarians or vegans.

I adopt an intersectional theoretical framework to help understand the multiple paths that could lead to self-identification as a vegan or vegetarian. Crenshaw (1991) originally introduced the term, *intersectionality* to describe the interconnected nature of social categorizations such as race, class, and gender as they apply to a given individual or group. These interconnected categorizations can thereby create overlapping and interdependent systems of discrimination or disadvantage. Intersectionality can help identify and explain the unique lived experiences of, for example, women of color and white women. While a white woman may face discrimination based on her sex (but not her race), a black woman may face discrimination based on her sex *and* race. The combination of race and sex creates a unique experience and treatment that is not simply the combination (e.g., additive) of race and sex.

Several scholars (Adams 2015; Joy 2018; Kemmerer 2011; Taylor 2017) have applied intersectionality to the treatment of animals. In principle, the oppression of animals could also be a function of intersecting categorizations that create unique treatment for those animals. For instance, Harper's *Sistah Vegan: Black Female Vegans Speak on Food, Identity, Health, and Society* (Harper 2010a) focuses on the similarities between the intersectional experiences of black women with those of animals (see also Aph Ko and Syl Ko 2017). The intersecting categories of black women can engender systematic oppressions that worsen the quality of life and opportunities for those black women. While white women and black men are discriminated

DOI: 10.4324/9781003039013-3

against only because they are members of those categories, being a black woman creates a unique, and in this case potentially destructive, experience that is not simply the sum of the discrimination experience by black people and women. Similarly, some animals occupy the intersection of different categories that creates a unique, and destructive, experience for them.

Intersecting categories can influence or inform oppressive systems and also help in efforts to dismantle, or even maintain, those systems (Harper 2010b). In the context of animals used as food, dietary habits and food production can connect to social justice issues. If we view this oppression through an intersectional lens, we can begin to see that unjust and immoral treatment of humans can also be applied to non-human animals. So, changing what we eat and following a vegan diet can be understood as a call for social justice in general (as an instance and recognition of the systemic injustices experienced by entities, regardless of species).

Food narratives can help shed light to the intersectional layers of vegetarianism and veganism. Listening to vegetarians and vegans share why they decided to challenge and change dietary habits and where on the path to veganism they position themselves illustrates how even vegans might not be purists and might still face a learning curve on their own food journey. Therefore, analyzing food narratives can lead to a storied activism that can support omnivores, flexitarians, vegetarians, and vegans in making ethically informed food choices.

The forgoing discussion suggests that there are many different ways that one could approach and start to identify as being vegan or vegetarian. As mentioned, some of these factors include notions of social justice. But other factors could include demographic features (e.g., income). Still others could involve an understanding of what animals used as food experience and a recognition that their interests are not being considered. Last, the sharing of food stories can complicate the often too narrow and simplified way we look at veganism. Not all vegans have turned 100% vegan in one day. Rather, veganism can be understood as the goal of a food journey and being on the journey itself has intrinsic value. If all of this is true, then there should be ways to detect those intersectional influences on people's food choices, and those intersectional influences should result in different, unique ways that one could become a vegan or vegetarian. To help provide some potential evidence for these intersectional relations as they relate to becoming vegan, I employed a mixed-methods set of empirical studies.

A Quantitative Study of Intersectionality and Vegetarianism

The primary goal of the survey-based quantitative study was to measure some commonly suggested factors that have previously been shown to be associated with adopting vegetarian lifestyles. To be clear, these are not the *only* potential paths that could impact views of vegetarianism, and they are not the only ways to measure different factors. Rather, the goal of the quantitative study was to develop a model of those factors that is consistent with hypotheses generated from intersectionality—namely, that there are likely to be multiple and unique ways to arrive at (and perceive) vegetarianism. Information on survey participants ($N = 152$) and survey material of the quantitative study can be found in Appendix.

Results and Discussion

Because of power issues and ease of analysis, I combined those who reported being vegans and vegetarians into a "vegetarian" category (Vegan $N = 32$, Vegetarian $N = 9$). There were significant positive correlations found between eating behaviors and conventional activism attitudes, the 4Ns, knowledge, political attitudes, and education level and income (see correlations in Table 3.1). Contrary to predictions, there was no reliable correlation with social justice attitudes or intentions, sex, or conscientiousness.

Only looking at the correlations among the variables does not really give a sense for intersectional relations. They only show whether *one* variable changes as a function of *another* variable. However, intersectional relations should be more complicated with multiple variables contributing to differential paths that result in choices to be a vegetarian.

To explore the potential intersectional factors associated with choices to be a vegetarian, a path model of factors that are likely inter-related in one's vegetarianism was developed. The hypothesized model along with standardized path coefficients is represented in Figure 3.1.

The observed data fit the hypothesized model χ^2 (12) = 13.44, p = .34, RMSEA = 0.03, 90% confidence interval = 0–.096, CFI = .99, and TLI = .99. All the direct effects represented in the model were significant ($p < .05$). All of the indirect effects leading to vegetarianism were significant except for the indirect effect of income (p = .08).

Table 3.1 Spearman's (non-parametric) correlations among the measured dependent variables * $p < .05$, ** $p < .01$ (1-tailed). Omnivores were coded as 1, vegetarians coded as 0.

	1	*2*	*3*	*4*	*5*	*6*	*7*	*8*	*9*	*10*
1. Omnivore	1									
2. Conventional	−.16*	1								
3. Attitudes	.09	.26**	1							
4. Intentions	−.09	.55**	.55**	1						
5. 4Ns	.65**	−.2*	−.15*	−.2**	1					
6. Knowledge	−.33**	−.02	.13	.11	−.5**	1				
7. Sex	.12	.05	.28**	.1	−.05	−.03	1			
8. Politics	.15*	−.58**	−.23**	−.38**	.28**	.08	0.3	1		
9. Post-grad	.16*	.21**	.05	.19**	.17*	−.18*	.03	−.17*	1	
10. Income	.14*	−.06	−.06	−.2**	−.21**	−.2**	−.06	.06	.36**	1
11. Conscientiousness	.04	−.06	.17*	.08	−.05	−.03	.11	.01	.03	.05

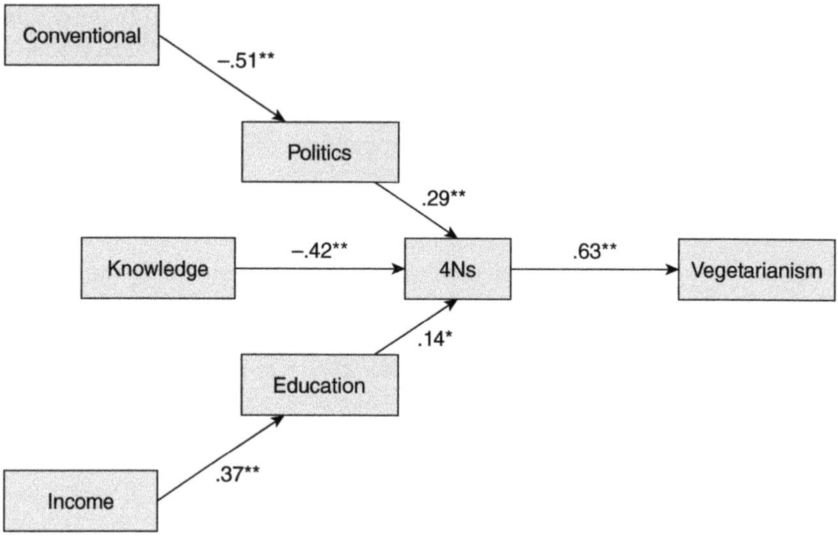

Figure 3.1 Path model of factors involved in vegetarianism. ** $p < .01$, * $p < .05$. Path coefficients represent standardized coefficients.

Given the results of the path model, we can draw some tentative conclusions about some of the different ways that one could adopt a vegetarian lifestyle. For instance, one potential way that one could adopt a vegetarian lifestyle is by being more politically liberal and being an activist. Those who are political activists and liberal might see the unjust and cruel treatment of animals as something that needs to be corrected. Alternatively, those who know more about animals used as food may be more likely to adopt a vegetarian lifestyle. Perhaps one reason why many people do not become vegetarian is that they simply do not know about the conditions that animals raised for food are in. Finally, those who have higher income and education may be more able to transition to a vegetarian lifestyle without much loss (e.g., they are able to afford expensive meat analogues). While I make no claims that these are the *only* ways that one could come to adopt a vegetarian lifestyle (e.g., there could be multiple layers of oppression that women and minorities experience that were not modeled), the model does provide some suggestive evidence that hypothesis generated from an intersectional framework are empirically supported.

"Because We Have Chosen a Life of Peace": A Qualitative Study of Vegan Food Narratives

The path model is suggestive that there could be multiple paths toward adopting a vegetarian or vegan lifestyle, largely by the different paths contributing to a reduction in attitudes that would justify using animals as food (i.e., the 4Ns).

Each pathway is suggestive that the route to becoming vegetarian or vegan is a process and does not necessarily happen in an instant. However, to provide more direct evidence from vegans themselves, I conducted a series of qualitative interviews to get a better sense of the reasons that vegetarians themselves offer for why they become vegetarian. As we will see, those justifications and reasons that the vegetarians offer echo the factors represented in the path model (i.e., paths to vegetarianism are a process and not a single event; the ability to be a vegan; knowing about animals used as food; and concern for general social justice). The path analysis gives us valuable information about the vegan identity but it is incomplete since participants had to decide on one rigid category (i.e., omnivore, vegetarian). Listening to food narratives allows for a more nuanced analysis because stories teach us more how the vegan identity can be seen as multi-layered and complex.

Participants

Out of 61 vegan/vegetarian survey participants, 16 vegans gave consent to be interviewed via Skype at a later date. Out of these participants, ten vegans responded to the emails that invited them to an interview. The participants' age group ranged from being in their 20s to being in their 60s. Two men and eight women were interviewed in the time span between May 5 and May 29, 2018. All of the participants lived in the United States. To protect the identity of the participants, they will be referred to as Participant#1–Participant#10.

Transcription and Interview Procedures

The ten interviews were semi-structured and ranged from 20 minutes to 45 minutes with the majority of the interviews in the 20-minute range. A few guiding questions served as content signposts to ensure that every interview covered the same main points and to keep participants on track. Each transcription was coded and analyzed using NVivo software. By coding qualitative data using nodes (i.e., references to the same topic), we can begin to understand in more depth the tendencies that result in being vegan. The following nodes were created in order to develop an understanding of vegan narratives and intersectional tendencies (see Table 3.2).

Moreover, a set of nodes was also used in order to understand what motivated participants to become and stay vegan and how they made the decision of leaving animal products off of their plates (see Table 3.3).

Last, a cluster of four nodes was used to look at some general attitudes of vegans (see Table 3.4).

Outcomes and Discussion

Three major strands become evident when looking at the interview data. First, and consistent with the path model above, *knowledge* played a

Table 3.2 Nodes used to code intersectional tendencies

Intersectionality: Statements that illustrate an awareness of the interconnectivity of systemic oppressive systems, such as disregarding the female body when it comes to reproductive rights.
Feminism: Statements that illustrate knowledge and an awareness of biases and discrimination women face, such as unequal pay or the sexualization of women's bodies.
Racism: Statements that illustrate knowledge and an awareness of linguistic and behavioral biases and discrimination based on race, such as comments about somebody based on their race.
Environmentalism: Statements that illustrate an interest or willingness to engage in service or activism that focuses on improving the environment, such as the idea of not driving a car.
Ableism: Statements that illustrate an awareness of the fact that the disabled receive treatment from institutions and society that others them form what is considered "normal."
Ageism: Statements that illustrate an awareness of the fact that older people receive treatment from institutions and society that others them from the rest, such as not being taken seriously because of one's age.
Speciesism: Statements that illustrate a sensitivity toward the idea that we treat animals according to their species, such as pets versus farm animals.
Health: Statements that share knowledge regarding the connections between diet and health, such as vegans tend to weigh less than omnivores.
Immigration: Statements that share a sensitivity regarding connections between how we treat animals and how we treat immigrants in the United States, such as locking refugee children in cages on the border.

Table 3.3 Nodes used to explore motivation to become vegan or vegetarian

Process: Statements that share knowledge or personal experiences about veganism taking longer than an initial Aha-moment, such as transitioning to veganism takes time.
Empathy: Statements that share an understanding of how animals feel, such as understanding that the violence farm animals experience causes physical and mental pain and stress.
Knowledge: Statements that share knowledge about food, factory farming, animal rights, and animal ethics, such as facts about slaughter and meat production.
Vegetarianism: Statements that share personal experiences and nutritional/ethical facts surrounding vegetarianism and veganism, such as a personal story of how to become vegetarian or vegan.
Veganism Then and Now: Statements that share knowledge about how veganism has changed over the years, such as the purist claim that argues how one can only be truly vegan if one never uses any animal products in any way.
Reducetarianism: Statements that discuss the idea of reducing one's intake of animal products, such as the lifestyle that condones animal products once a week or making deliberate choices to reduce meat and dairy in one's daily diet.

Table 3.4 Nodes used to categorize general attitudes

Individuality/ Identification: Statements that share how diet can shape and manifest one's individuality and/or identity, such as worldviews that show how one's veganism affects choices beyond food (clothes, medicine).
Activism: Statements that illustrate the willingness to serve and participate in animal or other activism, such as attending a protest.
Welfare: Statements that illustrate an interest and/or willingness to donate to causes (both for animals or humans), such as donating to PETA.
Lifestyle: Statements that allow insights into how vegans feel included or excluded from the mainstream lifestyle, such as being invited to a holiday party as the only vegan.

substantial role in people's self-reports of how they became and stayed vegetarian. The more interview participants learned about animal cognition and the realities of factory farming, the more likely the eliminated animal products from their diet. Second, interview participants also stressed how *process-driven* veganism as a lifestyle and as a movement is. While an immediate transition to veganism is possible and does happen, it is more realistic to assume that veganism does not happen overnight for everybody and that some simply need more time to adapt to the new lifestyle. Third, the interview data shares some insight into *intersectional perspectives* of vegans which can be helpful in developing a vegan activism that fights oppressive systems in general and not only the oppression of animals.

Knowledge

After transcribing the recorded interviews and uploading the transcriptions to NVivo, I used the nodes to categorize the interview data to look for qualitative pathways that could potentially explain veganism and some of the motivations behind veganism.

The word count calculator in NVivo indicates that *know* or cognates of *know*, such as *knowledge* or *knowing*, appear the most after the word, *vegan,* or cognates of *vegan*. In all ten interviews combined, the word *know* or cognates of it appear 205 times. The frequency of "know" should be taken with a grain of salt because fillers, such as "you know" are also included in this number. The next word of importance listed after *vegan*, *know*, and *like* is *thinking* or cognates of *thinking* which has been mentioned 196 times throughout the ten interviews. Thinking leads to knowledge and so it makes sense how these two words are closely connected and in the center of the word cloud besides veganism (see Table 3.5 for complete word counts).

Even though the word *know* is listed as one of the top word choices, it helps to look at the interviews more closely to see if knowledge/education truly plays a crucial role for vegans. After analyzing the ten interview transcripts, it becomes evident that all survey participants experienced an "Aha" moment or perception shift because all of them became vegan later on in life instead of

Table 3.5 The NVivo Word Calculation lists the frequency of words and their cognates in all ten interview transcriptions

Word	*Count*	*Weighted percentage*	*Similar words*
Vegan	282	2.57%	Vegan, veganism, vegans
Know	205	1.87%	Know, knowing, knows
Like	203	1.85%	Like, liked, likely
Thinking	196	1.79%	Think, thinking
Animals	179	1.63%	Animal, animals
People	155	1.41%	People
Really	150	1.37%	Really
Just	139	1.27%	Just
movement	109	0.99%	Movement, movements

being raised vegan by their parents. This Aha-moment was triggered in every participant, at least in part, by learning about the realities of factory farming or about animal ethics. The Aha-moment did not necessarily translate into direct action for most interviewees (see discussion of *process* below).

Since the members of the sample were generally more educated than the average American, it is not surprising that knowledge was a prominent factor in people's narratives. It makes sense that knowledge plays a vital role in challenging one's food habits because if one does not know about the injustices or treatment of animals (among other factors), it would be difficult to reliably change behaviors. Without knowing the relevant facts, one could consistently continue to use animals as food because of ignorance (Feltz and Feltz 2019). Adopting a vegan lifestyle can involve challenging one's own taste preferences, questioning cultural and societal norms, and learning new ways to prepare common dishes by using substitutes (along with the intersection of those different factors).

Participant #5, for instance, emphasized how reading John Robbins' *Food Revolution* "changed my life. And, uhm, I tried to read it a couple of times, but I wasn't really ready. You know, I wasn't. I wasn't personally ready to hear what the information had to tell me. And it was probably the third try and then I just devoured the book. And after that point it blew my mind. I said I can't do this, so I'm going to be vegan" (Participant #5). In this case, reading a book on animal ethics, nutrition, or veganism appeared to cause a shift in Participant #5's perceptions of animals as food—but only after Participant #5 was prepared to receive that information. An invitational approach to vegan activism helps people be more open to receiving messages instead of delaying their move toward veganism. Books, movies, or other information sources were explicitly also stated by Participant #1, 2, 9, and 10 as sources of information that changed their views on animals used as food. These results are consistent with other work that suggests that many vegetarians retrospectively report that information sources including feature-length documentaries, conversations with trusted individuals, online videos, social media posts, online articles or blogs, and books changed their minds about animals used as food (McCormick 2019).

Another common kind of knowledge mentioned concerned some of the realities of the dairy industry. These realities varied from mother cows being impregnated against their will (Participant #1) to dairy products being unhealthy (Participant #2). The participants did not go further into detail when it comes to the dairy industry. Moreover, scientific facts were mentioned when it comes to the dairy and egg industry, such as chickens don't have enough space in cages (Participant #4).

Process

The data analysis of eight of the interview transcriptions show how veganism is a process that takes time. All participants had been vegetarian by the time they considered veganism seriously, so vegetarianism can be understood as the first step people often take toward becoming vegan. But not only are people's food choices part of the process of veganism. The vegan movement itself has undergone change throughout recent times and has evolved in new ways.

Participant #10, a vegan for decades, goes on to share her opinion about the changes the vegan movement has been through and how it has in some ways, despite many advances, been its own enemy that created more obstacles than needed. The participant acknowledges the process the vegan movement in itself has been experiencing and how veganism has changed over the years. One main obstacle the vegan movement created is the idea of purity. Many vegans in online groups on social media publicly criticize and/or shame individuals who declare being vegan but then decide to consume fish or dairy again. This often happens to public figures who face the scrutiny of vegans when admitting they have eaten animal products or, what's even worse, if they get caught in the act eating fish at a restaurant (Arnold 2019). While not all vegans have these high demands, often loud voices in the movement seem to be voices who do not hold back their disdain and criticism when showing their disappointment. These voices unfortunately often turn curious people away from veganism because they either feel they cannot live a "pure" vegan lifestyle, or they feel being yelled at by the food police. Participant #10 shares how being vegan over 30 years ago was very different than it is now because of the emotions of purity:

> Now, there is real emotions of purity. It was very different when I became vegan. And there were no vegans, and nobody knew what vegan was. It's a very different world now than it was then. Then, vegans, I don't know. But now, people ask if a whole company is vegan. So, I would say it keeps changing, what it is to be vegan. So, that makes it a little trickier to say, was I vegan then? Well, certainly by my accounts I was. But [that was] by the accounts of the time. (Participant #10)

Participant #10 also discusses how eating eggs from one's own chickens that have been kept in an ethical manner is acceptable whereas "pure" vegans, according to her, would not condone that practice (Participant #10). However, the Vegan Society's definition is not that different from Participant #10's definition if we remind ourselves that humans can be considered animals, too. This way, humans are included in the definition of the Vegan Society's list of beings who should not be exploited or treated cruelly. Vegans who do not include humans in that list and therefore only focus on the suffering of animals misread this definition and miss how the definition in itself can lead to intersectional thinking.

The qualitative data of this study on the rhetoric of veganism illustrates that veganism takes many different forms and goes through a variety of stages and qualitative measures. It also seems that animal ethics might or might not be the beginning of a vegan narrative while animal ethics clearly plays a significant role after having been through one's personal stages of veganism. These stages can look differently for everybody and are highly influenced by one's own culture, support system, and belief system. What can be said about all but one interview (Participant #8) is that veganism does for them look like a slope or a learning curve where new lifestyle choices are made after more knowledge is acquired.

Experiencing an Aha-moment triggered by a documentary, a personal experience with an animal, or learning more about animal sentience or factory farming can be a powerful step in the food journey for an individual. It is important to note that becoming vegan is as important as being vegan since including individuals who try to make more ethical food choices can support their decision in adopting a vegan lifestyle. Looking at food narratives through a rhetorical lens that focuses on an invitational framework is beneficial in identifying this Aha-moment and first crucial step in becoming vegan and the vegan identity.

After experiencing the steps, and the perception change that ethical veganism evokes, it might not surprise how the belief system that comprises an ideology of a person can change throughout time as well. One of the first changes when turning toward veganism is an *empathy awakening*. Participants of the interviews mention a heightened sensitivity when it comes to the treatment of animals, and this sensitivity either had been dormant or deeply hidden in their subconscious. This can be compared to a stubborn ignorance because one is aware that a more direct engagement with the facts would bring along emotional engagement that one might simply not be ready to deal with. Participant #6 remembers how educating herself on what we eat by reading Jonathan Safran Foer's *Eating Animals* caused an emotional turmoil:

> I mean, I've always been an animal lover but like I've been, I really have been focused more on people than on animals like just my own work and I still am, I mean just like the way we treat animals now and just like the way we call it like, humane slaughter and things like that. However,

> the rhetoric is, it's just absolutely atrocious. I mean, it's like the stuff I can't, reading that Eating Animals book, that's the first and only book that I've ever read that I've cried, like I've never actually read a book and I cried, it's kind of weird but like, I actually had to put it down and I think a lot of people like [to] avoid looking at stuff like that because, so you have to, I mean, you can't avoid it, you know? We do so much to make ourselves comfortable and like, you know, and it's happening and it's really atrocious and something needs to be done because there's, you can't be like an ethical person whether or not you like animals or not just let that happen. It's just you know, it's the practices. It's disgusting. (Participant #6)

Participant #6 shows how we can expand our moral horizon so as to overcome speciesism. When Singer popularized the concept of speciesism, he did so through an intersectional perspective by comparing it with racism and sexism. Singer (1975) defines speciesism as "a prejudice or attitude of bias toward the interests of members of one's own species and against those of members of other species." Learning about the practices and the systematic oppression of farmed animals can be a powerful step toward expanding one's moral horizon and veganism.

The process of becoming vegan does not end by being vegan. Even vegans expand their perspective after eliminating animal products from their diet. Participant #9, for example, starts to include her natural world into her moral horizon in a way she had not done so years ago. One could argue that the environmental movement is simply getting stronger as more information is spread about how human behavior affects the Earth. One could also argue that participant #9 clearly had a strong emotional bond with the hiking trails in and around her home and that she maybe was per default emotionally more invested when the fires happened. Still, Participant #9 is aware of "little changes to how [she] perceive[s] the world" which draws attention to the perception shift that happens after becoming vegan (Figure 3.2).

Intersectionality

The analysis of the narratives also revealed a tendency to include oppression of other marginalized groups. It is one thing to say one cares about marginalized groups. It is another thing to do something about it. While Participant #4 is more hesitant in seeing the similarities between oppressive systems and marginalized groups in general, all other participants would, to an extent, include veganism into the same category they put other social justice movements. Participant #1 and Participant #8 go back to their childhood to explore their attitudes about oppression. Both felt early on how they had a sensitivity for "the underdog" and were keenly aware of injustice (Participant #8). This sensitivity stuck with them into adulthood and might have influenced their decision to go vegan. Participant #8 grew up

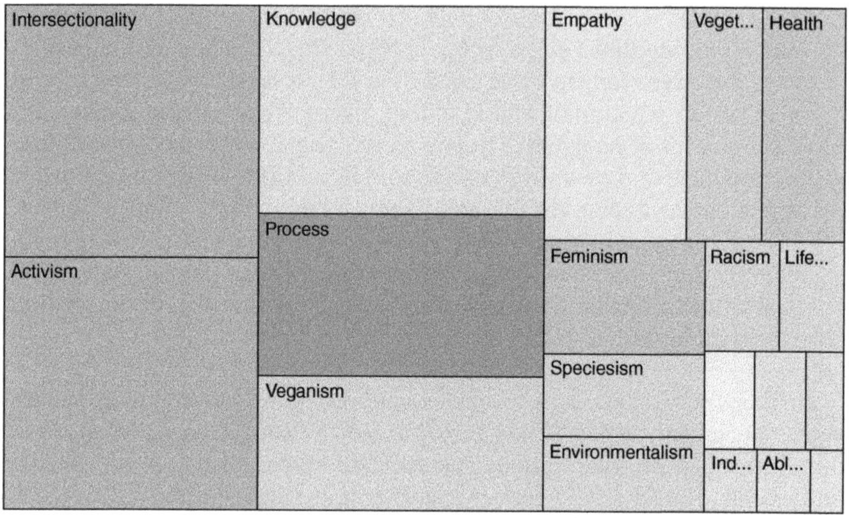

Figure 3.2 The NVivo Treemap is a visualization of how often the above listed nodes were used in all ten interview transcriptions. More frequently used nodes like intersectionality or knowledge are represented by taking up more space than less frequently used nodes like health or individuality. The empty boxes represent the nodes welfare, immigration, veganism then and now, and reducetarianism.

with several older siblings and learnt early on to become an ally for those who were considered less strong:

> I'm the youngest of seven children in a volatile family and I had to learn to defend myself and I always identified with the underdog. And from a tiny age, like 3 or 4, I began rescuing animals, like children who would be hurting animals or my mother about to hurt a mouse and I would go between them. And I have been active in the feminist movement. My very first political action or activism was when I was just 13 or 14, I guess, and I was involved in an environmental action at Cambridge mass where students and people were trying to save about 100 sycamores that were lined along [...] River in Cambridge. That was my very first movement, my very first action with, you know, adults, and it was heavy. Now, we won! The trees are still there. We saved them. So, I thought, it could be done, working together. Injustices can be dealt with. So, I marched in the civil rights movement in Massachusetts where the schools were very segregated and the whole towns were segregated and when the federal law required desegregation, the mayors of a couple of towns in Massachusetts actually marched to stop the busses, hauling the children, black children, into their towns. [...] I'm not a Catholic, I am an atheist, but I was brought up in an Irish Catholic family. So yes, I consider all these justice movements intertwined. (Participant #8)

When asked if there was anything else she would like to share at the end of Interview #2, Participant #2 made the following statement:

> [B]ecause we have chosen this truly; a life of peace because we are saying we are not going to kill animals, we are not having a baby calf cry for its mother [...] kind of like what you mentioned about all the racism and the other -isms and all that. All the other movements. It seems like once you, once you see that; the hate of how we are treating these animals; it seems like we've already, we've already. I don't know. I don't know vegans that are racist, you know? I don't. (Participant #2)

Participant #2 continues to wonder if the data I collected shows how vegans might be more peaceful in general, not only peaceful toward non-human animals (Participant#2). By pointing out how vegans tend to be less violent, in her opinion, she acknowledges how veganism extends to human suffering and the empathy and sensitivity that can come along with the before-and-after-effect of a vegan lifestyle directly impacts other marginalized groups, as well.

Veganism and vegan activism in particular take many different forms. Ethical veganism can be focused on animal rights alone. A person can identify with ethical veganism without showing much empathy or understanding of other forms of oppression. However, if intersectional thinking becomes a vital part of ethical veganism, then other social justice movements can be included on the moral horizon and vegan activism also becomes part of the feminist movement or the LGBTQ+ movement. If social justice is incorporated into vegan activism, an intersectional vegan activism can contribute to veganism being less isolated and more integrated as a lifestyle that opposes all forms of oppression—and one that is more likely to reflect the reality that there are many paths to becoming a vegetarian or vegan. This kind of activism can not only advance the vegan movement but might advance other social justice movements as well. The role of an ally is crucial in making the vegan movement more inclusive. Being an ally to vegans, as a non-vegan, helps everybody understand the food preferences of others better and fosters better communication. However, vegans need to be allies to other social justice movements as well and support other oppressed groups beyond animals. Recruiting allies for a movement is a two-way-street. Offering to be an ally in other social justice movements and also welcoming non-vegans as allies to be a vital part of the vegan movement allows social justice to become part of ethical veganism which in turn can lead to intersectional veganism.

Conclusion

Living a vegan lifestyle can be traced back to a multitude of factors and attitudes. My data show that some factors intersect with one another and should not be viewed as separate strands or pathways. Veganism this way can be understood a multi-layered belief system and lifestyle that could also benefit from a kind of activism that focuses on this multi-layeredness rather than siloed aspects of veganism or "one-size fits all" approaches to activism. The sharing of food narratives can be one channel for this storied activism since storytelling allows the listener to become more familiar with nuanced attitudes and beliefs.

Familiarizing ourselves with the personal, lived experiences of vegans helps us see how knowledge, process, and intersectionality play a substantial role. Through learning about veganism, individuals can make informed decisions based on research and facts. This knowledge base changes with time and with the exposure of new vegan products which also emphasizes how process and knowledge are intrinsically linked with one another. Granted, some vegans abstain from consuming animal products after watching one documentary or after exposing themselves to graphic factory farming coverage on YouTube, but many vegans turn to veganism more gradually. This slow process toward veganism is a more realistic approach to a lifestyle change that asks a deep engagement with products, labels, and lifestyle questions in general.

If we focus rather on veganism as a process instead of a purist lifestyle, we acknowledge many individuals making decisions toward becoming vegan and support them in their food journey. Looking at veganism only as a strict diet with no fallbacks limits our rhetorical understanding of veganism. A narrative approach helps understand these nuances better and a rhetorical understanding of the vegan food journey can this way be developed in a more meaningful way. This way, quantitative research allows us to see certain food trends while a qualitative framework helps us analyze why individuals make the food choices they make as they move toward a vegan lifestyle.

An invitational framework that stresses mutual openness and equality between all parties of the rhetorical situation helps in illustrating the layers of storied activism that leads to becoming vegan. People who are on the path toward veganism can this way be identified more clearly and rather than judging a person's non-vegan food choices, an empathy-fostering invitational framework can be more inclusive and supportive in people making better decisions in the future.

Listening to the stories of vegetarians, vegans, reducetarians, and omnivores allows us to connect with individuals because others might experience similar struggles and triumphs in our quest to learn about the best way to eat. At the same time, the exposure to food narratives separates us from others because we don't share the same believes and attitudes about animal rights, food ethics, and the culturally based choices we make regarding food. Differing from others helps us question our personal food narrative and we have the opportunity to change and help not just ourselves live a better and healthier life but we also help animals and the planet.

Appendix: Quantitative Survey Materials

Participants

One hundred and fifty-two participants recruited using snowball sampling techniques. Sixty-one reported being vegetarian or vegan (40%) (2 people did not indicate dietary preference). A large percentage of the sample identified as female ($N = 116$, 76%, 5 people did not indicate sex). The average age was 40.96 ($SD = 12.96$) ranging from 18 to 78. Sixty-percent of the sample ($N = 91$, 1 declined to respond) had a postgraduate degree, and 47% ($N = 72$, 16 declined to respond) had a household income of more than $74,999 a year. There was also a higher than the U.S. average number of atheists or agnostics in the survey, 57% ($N = 85$, 12 declined to respond).

Materials

After providing informed consent, participants were presented with the following instruments. Theoretically, each instrument was selected because the literature on intersectionality and the broader psychology of vegetarianism suggests that these factors should be related to different paths that one could come to vegetarianism by.

The Activism Orientation Scale

The Activism Orientation Scale (Corning and Myers 2002) is a general measure of one's attitudes toward activism. The 35-item scale consists of two subscales measuring conventional activism (e.g., whether one would be encouraging a friend to be an activist) and a high-risk activism subscale (e.g., whether one would engage in a physical confrontation at a rally). It was predicted that conventional activism would be negatively associated with being an omnivore.

The Social Justice Scale

The Social Justice Scale (Torres-Harding et al. 2012) is a 24-item measure of people's attitudes toward social justice. The scale consists of 4 subscales that measure one's attitudes about social justice (e.g., "I believe that it is important to talk to others about societal systems of power, privilege, and oppression"), one's control concerning social justice issues (e.g., "I believe that it is important to allow others to have meaningful input into decisions affecting their lives"), what one views as norms that govern social justice (e.g., "Other people around me are supportive of efforts that promote social justice"), and finally, intentions concerning social justice issues (e.g., "In the future, I will do my best to ensure that all individuals and groups have a chance to speak and be heard"). It was predicted that attitudes and intentions about social justice would be negatively associated with being an omnivore.

The 4Ns

The 4Ns (Piazza et al. 2015) is a measure of how Natural, Nice, Normal, and Necessary eating animals is (e.g., "Our human ancestors ate meat all the time"). It was predicted that the 4Ns would be positively associated with being an omnivore.

The Knowledge of Animals as Food Scale

The Knowledge of Animals as Food Scale (Feltz and Feltz in press) is a measure of objective knowledge about animals used as food (e.g., "Most chickens used for food and eggs live in open spaces"). It was predicted that the Knowledge of Animals as Food would be negatively correlated with being an omnivore.

The Ten Item Personality Inventory

The Ten Item Personality Inventory (Gosling et al. 2003) is a brief measure of the Big Five Factor model of personality (extraversion, agreeableness, conscientiousness, emotional stability, openness to experience). It was predicted that conscientiousness would be negatively related to being an omnivore.

Demographics

Basic demographic information was gathered. It was predicted that being female, being politically liberal, having a graduate degree, and being wealthier would be negatively associated with being an omnivore. It was also predicted that those of minority groups would be more likely to be vegetarian. But, given the constraints of the snowball sampling technique, not enough minorities were included in the sample for meaningful analyses to be performed.

Bibliography

Adams, C. J. (2015). *The sexual politics of meat: A feminist-vegetarian critical theory* (Bloomsbury Revelations Edition). New York, NY: Bloomsbury.

Arnold, A. (2019). Vegan blogger Rawvana accused of eating fish. *The Cut*. https://www.thecut.com/2019/03/vegan-blogger-rawvana-accused-of-eating-fish.html

Corning, A. F. & Myers, D. J. (2002). Individual orientation toward engagement in social action. *Political Psychology, 23* (4), 703–729.

Crenshaw, K. (1991). Mapping the margins: Intersectionality, identity, politics, and violence against women of color. *Stanford Law Review, 43* (6), 1241–1299.

Feltz, S. & Feltz, A. (2019). The knowledge of animals as food scale. *Human-Animal Interaction Bulletin, 7* (2), 19–45.

Gosling, S. D., Rentfrow, P. J., & Swann, W. B. (2003). A very brief measure of the big-five personality domains. *Journal of Research in Personality, 37*(6), 504–528. doi: 10.1016/S0092-6566(03)00046-1

Harper, A. B. (Ed.) (2010a). *Sistah vegan: Black female vegans speak on food, identity, health, and society.* New York, NY: Lantern Books.

Harper, A. B. (Ed.) (2010b). *Sistah vegan: Black female vegans speak on food, identity, health, and society* (p. xix). New York, NY: Lantern Books.

Joy, M. (2018). *Beyond beliefs: A guide to improving relationships and communication for vegans, vegetarians, and meat eaters.* Brooklyn, NY: Lantern Books.

Kemmerer, L. (2011). *Sister species: Women, animals and social justice.* Chicago, IL: University of Illinois Press.

Ko, A. & Ko, S. (2017). *Aphro-ism: Essays on pop culture, feminism, and black veganism from two sisters.* New York, NY: Lantern Books.

McCormick, B. (2019). The 2019 global vegan survey. *VOMAD.* https://vomadlife.com/blogs/news/why-people-go-vegan-2019-global-survey-results?fbclid=IwAR1p46dR2Bhz8Drg_QwD32Xnf0iCaZxBddUTBWFPvZmO1INzEYMVgL1UIlg.

Piazza, J., M., R., Loughtan, S., Luong, M., Kulik, J., Watkins, H., & Seigerman, M. (2015). Rationalizing meat consumption. The 4Ns. *Appetite, 91,* 114–128.

Singer, P. (1975). *Animal liberation. A new ethics for our treatment of animals* (p. 7). New York, NY: Avon Books.

Taylor, S. (2017). *Beasts of burden: Animal and disability liberation.* New York, NY: The New Press.

Torres-Harding, S. R., Siers, B., & Olson, B. D. (2012). Development and psychometric evaluation of the social justice scale. *American Journal of Community Psychology, 50,* 77–88.

4 The Bounded Discourse of the Whole Foods Plant-Based Movement

D. R. Hammontree

Introduction: From Veganism to Plant-Based Solutions

Marketed with a quote from the late movie critic Roger Ebert (2011) as "A film that could save your life" the 2011 documentary *Forks Over Knives* (Fulkerson 2011) was influential in bringing the vegan or Whole Foods Plant-Based (WFPB) diet debate into popular public discourse. The documentary featured various WFPB health advocates with credentials ranging from traditional MDs and PhDs, including Neal Barnard and T. Colin Campbell. The framing of the narrative was told from the vantage point of Campbell, a researcher working out of Cornell University, and Caldwell Esselstyn, a heart surgeon based at the Cleveland Clinic. Both find that their research interests coincide and argue for the benefits of a diet without non-human animal products. The claims made in the documentary, that a WFPB diet could reverse some diseases and even cancer, were bold and controversial.

Forks Over Knives also gives time to alternative health practitioners including chiropractor Doug Lisle and naturopath Pam Popper, both representing strands of lifestyle movement health advocacy. Although the vegetarian movement has a history with the alternative medicine, this alliance of WFPB advocates from two otherwise conflicting fields leads to an intersectional, yet bounded discourse community. A strange coupling, indeed, of two groups has created a united front advocating for a plant-based diet.

This chapter focuses on the background and discourse of WFPB diet advocates of veganism, highlighting how those "plant-based" discourses grounded in an ethos of healthism (Crawford 1980) may do a disservice to the larger, more socially minded vegan movement. The rhetorical framing of the internal and external discourses of the WFPB health movement demonstrates an attempt to reach a mainstream audience. But in reaching a wider audience, the WFPB movement distances itself from the political and ideological strands of ethical and environmental veganism. Further, the WFPB movement takes on the mantle of an outlier, appealing to an anti-elitist ethos as a means of setting itself up against "Big Medicine." The relationship between experts from traditional medicine and those from the alternative health communities also crosscuts into the WFPB community's

DOI: 10.4324/9781003039013-4

internal discourse. Wider debates in the WFPB community too often veer away from veganism and into diet, where Atkins or Paleo become argumentative foils, rather than wider ethical and environmental concerns.

The problem to explore begins with what amounts to a bounded discourse community. Discourse communities, as theorized by writing studies scholars, by their very nature can be seen as "constraining system[s]" (Porter 1992, 106). And although Porter's poststructuralist approach defining a discourse community is a bit "messy" (Evans 1998, 2) in its fluidity, the intention is to use this model to complement a wider theory of a *bounded* discourse community that narrows the concept and branches off from Simon's (1957) psychological model of "bounded rationalities" wherein "human behavior can always be explained in terms of pursuit of self-interest subject to the perceived opportunities" (Basov 2016, 3). WFPB health advocates create a bounded in-group identity that sets itself apart from both a wider vegan community as well as mainstream nutrition and medicine. The trouble occurs when these bounded discourse communities take up a defensive posture that closes down the potential for a wider discourse.

There is a push and pull between the individualized focus on health and the larger collective needs that may be addressed by veganism: a worry that the internal discourse of the WFPB group may be drifting away from the larger drives and external appeals of the movement. The bounded discourse community breaks off on its own when the discussion turns to only diet. Then, appeals of ethos of individualism takes hold to combat the impression of "moralizing or proselytizing" (Wilson, Weatherall, and Butler 2004, 570) from the animal welfare or environmentalism camps. Motivational reasoning becomes muddled in the discourse.

Motivational Overlap

There are three main and overlapping motivations for individual vegans: non-human animal welfare, environmental concerns, and health (Cooney 2014, 139–53). Individuals practicing a health-based veganism may find themselves more open to empathy toward non-human animals (Singer and Mason 2006), while the ethical and environmental concerns may influence how others come to understand the role that non-processed, plant-based foods may play in one's health and lifestyle. As well, participation in a vegan practice has the potential to help address overwhelming feelings of powerlessness many feel about the global climate crisis.

Laura Wright's *The Vegan Studies Project* (2015) cautions that "[a]s veganism more fully enters mainstream discourse, it remains to be seen how the tensions between animal advocacy and dietary proscriptions based on health and determined by sociocultural factors will shape the next discursive manifestation of veganism" (9). These tensions may be seen in an exploration of the rhetorical appeals and ideological framework of some members of the WFPB discourse community. First, there is a feeling of anxiety and insecurity on the part of the voices in the WFPB diet movement that make for an overall defensiveness—they feel shunned by mainstream medicine and, in turn, find alliances with alternative gurus. As well, many from the mainstream medical community run the risk of falling into a sectarian entrenchment that threatens to alienate them not only from mainstream medicine but also from the wider vegan movement.

Herbivore lifestyles have always been marginalized in a US cultural context, and have been associated with communal life or a pacifistic world view that gives more attention to non-human animal welfare. The concept of ethical veganism and vegetarianism is "both as an identity category and as a practice" (Wright 2015, 1) for animal advocacy while pushing against larger societal and cultural traditions. Wright also points out that today, "veganism has found a foothold as a mainstream practice [and, to] make veganism more accessible to a broader group of people, it has been alienated from its more radical animal-advocacy roots" (2015, 9) shifting more toward lifestyle. Adjacent and often allied to the ethical strand of vegetarianism was a health-focused strand of the practice, historically associated with alternative health practices and an anti-establishment, anti-elitist ethos that sees the medical establishment as entrenched in larger financial and political or simply professional interests (Iacobbo and Iacobbo 2004); we refer to this as alternative or naturopathic health care.

What is unique in this wider community is that the MDs in this group share the same space with advocates who carry alternative health credentials. The discourse community finds itself comprised of conflicting philosophies of medicine (the mainstream and the alternative), creating a credibility problem in how WFPB advocates may be received by a general audience. How credibility is established and maintained complicates the discourse on WFPB health movements.

The roots of this conflict are historical. In the 19th-century United States, as what had once been a traditional means of holistic healing gave way to an age of professionalism, credentialism gradually led to a social transformation of medicine in (see Starr 2017). With credentialism came a move toward professional legitimacy. "Expert authority" here became another means of legitimization (Van Leeuwen 2008, 107). Those practicing with traditional remedies, of which a vegetarian diet was but one, were pushed to the side as nutrition experts aligned with mainstreamed medicine also focused on narrow approaches rather that holistic views.

In the history of vegetarian health practitioners, cultural pressure remains an influence when it comes to the role of nutrition advocates and their relationship with the established medical community. But while all seem to be in the same place on general nutritional and diet advice, an underlying professional ethos from an "authority of tradition" (Van Leeuwen 2008, 108) may be a separating factor. The largest tie between various forms of the WFPB diet movement comes in the resistance to the medical establishment and in turn places the movement as outside the mainstream. Indeed, the vegetarian movement has a thorny history within health and medicine in the United States (see Spencer 1995; Iacobbo and Iacobbo 2004; Starr 2017; Weisse 2020). Weisse (2020) charts this history and sees "orthodox medicine and alternative medicine [...] opposed, as were the food industry and the natural health food movement" (28). But medically trained, plant-based MDs and researchers share many of the frustrations of the alternative medicine community when it comes to the influence of the medical establishment and food industry.

Three Voices

The historical frameworks above set a stage for an overview of a few well-known personalities in WFPB health advocacy (all featured in *Forks over Knives*) cutting across ideological boundaries yet comprising a bounded discourse community. Although some of the voices below acknowledge the ethical and environmentalist threads of veganism, others are focused on more libertarian concepts of freedom and choice. All share a certain amount of alienation toward the medical establishment rather than appealing to animal welfare and environmentalism. The outlier status becomes shared trait, and, beyond giving general dietary advice, the combined arguments are grounded in the critique of bureaucratic politics of the US medical establishment, and, in turn, the federal government and certain business interest.

The advocates below were selected because of the public exposure received after to appearing in *Forks over Knives*. All four have all advocated the advantages of a WFPB diet. Further reading of their work introduces the general reader to the internal discourse of this community and its overall frustrations with other diets (low fat, anti-sugar, Paleo, Keto, or Atkins) and the advocates of those diets. One prominent WFPB advocate from the mainstream medical community featured in *Forks over Knives* is T. Colin Campbell. Alternative health practitioners coming out of chiropractic and naturopathy are Doug Lisle and Pamela Popper.

T. Colin Campbell

T. Colin Campbell is featured predominately in *Forks Over Knives* (Fulkerson 2011) and ranks high as a credible WFPB diet advocate thanks to his work *The China Study* (2005, 2016). Campbell's credentials are as a researcher with a specialization in nutritional biochemistry and as Professor Emeritus of Nutritional Biochemistry at Cornell University (Campbell and Jacobson 2013, 293). He is a leader of WFPB diet movement focusing on non-processed, plant-based food choices. *The China Study* is co-written with Thomas M. Campbell, II, who by the revised edition of the work became "a practicing, board-certified family physician" (2016, xiv).

The research in *The China Study* came out of population research initiated by the Chinese Government in the early 1970s (2016, 59). Gyorgy Scrinis points out, however, that *The China Study* "is only loosely based on the large epidemiological study of the same name" (2013, 182). The Campbells' analysis indicated that cancer rates in China were "geographically localized" and that rates were hypothesized to be "due to environmental/lifestyle factors and not genetics" (2016, 60). The remainder of the work makes the larger case that non-human animal-heavy "diets of affluence" are the predominant causes of the rise in cancer rates (2016, 65–7). Diets high in animal protein increase blood cholesterol, female hormone levels, and cancer rates (2016, 76). The Campbells point to a corrupting influence in science (249–302), government (303–20), "Big Medicine" (321–44), and academia (345–56) as reasons why research into non-human animal-based diets is not progressing. The overall argument comes down to the power and influence of the animal agriculture industry and other like-minded interest groups in maintaining the interest of the status quo.

In *Whole* (2013), Campbell, along with Howard Jacobson center their arguments on the effects of "reductionism," a working scientific and clinical paradigm that focuses on the pieces over the whole (2013, 288). Reductionism leads to a fixation on specializations that weaken appreciation of the complexity of interrelatedness of the body with nature: a similar pattern seen with sociology, psychology, or the myriad of positions that find it easy to dismiss any argument along pluralistic or larger systematic lines. Those funding and potential conflicts of interest are routinely considered when making decisions about the ethics of proposed research as a basic practice of institutional review boards is also a concern.

Douglas Lisle

Douglas Lisle holds a PhD in Clinical Psychology from the University of Virginia and was a Lecturer in Psychology at Stanford University. Lisle was also interviewed for the documentary *Forks Over Knives* (Fulkerson 2011). Lisle (along with chiropractor Alan Goldhamer) authored *The Pleasure Trap: Mastering the Hidden Force that Undermines Health & Happiness* (2003). The first four chapters of this book are a critique of the medical establishment, with an emphasis on understanding "health and happiness" (2003, 3) while advocating for "a better way" (2003, 36) before surgery options. Lisle and Goldhamer posit that the public's assumption that "ingenious medical procedures are likely to save them from any disease process that they may have to face" (2003, 37) contributes to the lack of comprehensive understanding of well-being. Relying on medical procedures for preventable conditions is part of this "pleasure trap." The overall approach is one of lifestyle and well-being. The authors' emphasis is on individual happiness, focusing on "mood systems work[ing] in conjunction with the pleasure-seeking and pain-avoidance systems [...] encourag[ing] effective behavior" (Lisle and Goldhamer 2003, 13). Arguments and criticism used through this book focus on the larger practices and public perceptions of the mainstream medical community. The two sidestep larger environmental and climate emergency issues for an "environment within" (2003, 129) approach, focusing on how individuals can gain a greater hold on emotional environments. In turn, outreach toward potential vegans motivated by animal rights or global climate concerns is not addressed in Lisle and Goldhamer (2003).

Pamela Popper

Pamela A. Popper exudes a sharp air of confidence and credibility in her interviews in the *Forks over Knives* documentary (Fulkerson 2011). Her book, *Food Over Medicine* (2013) is written in the form of an interview with screenwriter Glen Merzer. The work covers many commonsense ideas of nutrition, diet, and lifestyle. Chapters cover general plant-based diet arguments, causes of diseases, and eating plans, specifically that of Popper's Wellness Forum (2013, 26).

Popper's biographic blub describes her as a naturopath and "an internationally recognized expert on nutrition, medicine, and health" serving as "Executive Director of the Wellness Forum" (2013, 239), a Cleveland-based alternative health center. Popper serves on the steering committee for Barnard's PCRM and "is part of [...] Campbell's teaching team at eCornell, teaching part of a certification course on plant-based nutrition" (2013, 240). Popper's use of PhD and ND (naturopathic doctor) suffixes place her on par with others in the medical community and is an understandable means of establishing her ethos. The specifics about her degrees (what specialization and institution) however, are not revealed anywhere in the work, nor does any information about her degrees appear to be available at her personal or Wellness Forum websites. A 2005 article from *Cleveland Scene* reveals: "she's a naturopathic doctor and received her degree from Clayton College of Natural Health, a correspondence school in Birmingham, Alabama" (Maag 2005).

Topics in *Food Over Medicine* lead readers to a larger discussion about the governmental institutions and structures that influence dietary advice, including the role of the FDA, USDA, and others in the food industry (2013, 136–54). Popper argues that "the further away you get from doctors, the better off you're going to be, in most cases" (Popper and Merzer 2013, 156). The larger argument within is that most aliments could be "avoided [by] eating the right diet [and not] taking [...] dreadful pills" (Popper and Merzer 2013, 160). This is an argument informed by her personal experience and patient anecdotes. Merzer, interviewing Popper, addresses "education, clean energy, infrastructure, [and] the environment," i.e., "the real market cost of a cheeseburger" (Popper and Merzer 2013, 222). The overall work is grounded in a lay discussion of the economics and sociology of healthcare rather than non-human animal activism or environmentalism.

Legitimation Anxiety in a Bounded Reality

Forks Over Knives features the influencers above, degreed, or otherwise, promoting a lifestyle; the MDs lend legitimacy to the overall group. Now the questions of ethos as well as influence and credibility arise in a status-driven culture where credibility is hardly the determining factor in influence making. The unconscious ethos of a bounded discourse community of WFPB health professionals is predicated on an anxiety over criticism about credibility, either professional or research-oriented that comes out of making grand claims of a disease-reversing diet. A worry about being dismissed by the mainstream influences the rhetorical decisions of WFPB health experts. When credentials are questioned, so is the authority to make claims.

A discursive, rhetorical lens may best explore how ideological attitudes affect the continuing discussion about the advantages of a WFPB health. Basic to any discourse analysis is understanding what the community may believe "consciously or unconsciously [...] in order to make deep sense of what they are saying" (Gee 2005, 87). An audience grants legitimacy based on the authority it gives the expert, including personal authority, expert authority, and role model authority (Van Leeuwen 2008, 106–8). Van Leeuwen breaks down various forms of legitimation simply enough with personal authority coming out of upbringing, expert authority from credentialed professionalism, and role model authority from "advertising and lifestyle media" wherein celebrities gain legitimation (2008, 107–8).

How the public perceives what is authoritative, legitimate, credible, or otherwise is a thorny issue. In mainstream discourse, credentials alone don't influence the perception of legitimacy. Social influence makes for credibility more than credentials. Starr's history of American medicine, for example, informed heavily by Max Weber, sees that medical "authority spills over its clinical boundaries into arenas of moral and political action" (2017, 5). Weber (1978) and, in turn, Habermas (1975, 212–16) while not directly addressing rhetoric, both look at legitimacy as a cultural practice predicated on connections that are influenced by experience (Perelman 1982, 100). In ideals times, the general public places its trust in professionalism which, in turn, establishes credibility and the influence that comes with it (Perelman 1982, 31). The credential process along with established research expectations is, ideally, designed maintain legitimacy in the public mind while making itself open for any possible paradigm shifts (Kuhn 1996).

Such epistemological thorniness plays out in public consumption of medical and diet advice. Often the cultural influence of the speaker is far more influential than expert credentials. There is a hierarchy that makes for a defensive strategy among outlier communities. We can swap-out the term "authority" as a rhetorical concept with "legitimacy" in the passage below by George Kennedy:

> Authority from the position in a hierarchy is a powerful means of persuasion that is brought to bear on the rhetorical situation, sometimes without any specific reference to it in the words that are spoken or written. The presence of the speaker or the name of the author is sufficient [and] often persons in these positions have no need to supply reasons for their pronouncements to be effective.
>
> (1998, 42)

Outliers who can't gain influence in their own fields often find it in a public hungry for answers. While individuals would most likely take medical advice from personal physicians, public discourse on medical matters often finds experts of mixed backgrounds, some that may not be related to a given field or topic. Here legitimacy is a cultural chicken or the egg as brought on by a sense of Van Leeuwen's idea that personal authority that goes unquestioned by exposure becomes entwined with a perception of expert authority (2008, 106). Those personalized experts then become influential "role models." Legitimation feedback is thus the exposure that creates credibility or the credibility that such exposure creates.

The credentials of the influential figures interviewed in *Forks over Knives* reveal a conscious effort to build a professional ethos or authority—one that allows a willing audience to be more receptive to their arguments. Yet, by offering quick solutions to complex problems the "alternative" practitioners without the established credentials are often dismissively as "gurus" (Weisse 2020) preying on "seekers" (Cialdini 2009, 56). While individuals have less trustworthy credentials than others all are working to build "scientific" legitimacy of sort, in offering vegetarianism as a very alluring solution to a myriad of social and individual problems. But quick fixes are attractive *because* non-complex answers are offered to a public often left alienated by scientific practices (that by design need to distance themselves from their subjects). Yet, once this community takes shape, we come back to the danger of a bounded discourse wherein influencers get so caught up in their own echo chambers that criticism becomes their own ideological underpinnings. Traditional notions of credibility get lost in the shuffle.

Applying Perelman's principles of "choice, presence, and presentation" (Perelman 1982, 33) gets at value judgments and helps frame an exploration of credibility and legitimation introduced above. Falling under "choice" is the argument that "[e]very argument implies a preliminary selection of facts and values" leading to a "[c]hoice of elements, of a mode of description and presentation, judgment of value or importance" (Perelman 1982, 34). Inferring from Perelman's basic outline sees value judgments serving as the internal reasonings that attract individuals to the movement and help to maintain the cultural practice of health-based veganism. Further, the "presence" in the form of what is singled out for presentation further "acts directly upon our sensibility" (Perelman 1982, 35). *Forks over Knives* works from the expertise of the MDs while the emotional appeals and defensive appeals come from those practitioners outside the medical establishment. What these two sides of the WFPB movement have in common is their engagement with the established (i.e., mainstream) attitudes and politics surrounding how we consume and use non-human animals. The rhetorical moves in *Forks over Knives* work to ground those voices in traditional medicine, not joined by the naturopaths and other non-traditional voices, pushing back against conventions as soldiers in a greater struggle against Big Pharma, Big Food, Big Medicine, etc.

By speaking to a health conscious public receptive to a healthy lifestyle message, the WFPB diet advocates find themselves captive to the pressures of role model authority. Yet, there is a fear that aligning a health message to a wider social movement (animal welfare and environmentalism) would drive away a general audience. An ideological understanding about audience is essential to develop a rhetorical framework of public debate where, too often, "[e]very analogy highlights certain relationships and leaves others in the shadows" (Perelman 1982, 119). Perelman's shadows are evident in the WFPB community wherein professional anxieties lead to ideological entrenchment in a bounded discourse community. The rhetorical choice of grounding appeals as lifestyle promotion presents is frustrating to those who view veganism as a wider social justice movement rather than a diet. An ideological crossover places the vegan health advocates in one camp, while animal welfare and environmentalist vegans advocate from more leftist orientations.

Public Perception of Whole Foods Plant-Based Diets

Presently, the alternative health community takes on its traditional role of individualized attention and care for a larger percentage of individuals that may have lost faith in the current health system. Charges of "elitism and personalization" (Gregory and Miller 1998, 113) are the alternative health advocates reactionary pushback toward established medical professionals. Yet, in the world of WFPB, the alternative medicine practitioners (homeopathic, chiropractic, naturopathic) are often endorsed by the MDs or scientific researchers, creating a united front to a public that may not realize the professional differences between the two groups.

The professional training of practitioners like Campbell and Barnard make for credible ethos to point out corporate and commercial influences on accreditation bodies that Campbell targets, such as the Academy of Nutrition and Dietetics (AND) (2013, 271). But it is due to this very criticism of established institutions that Campbell and Popper find themselves allied. Campbell writes, "My friend and colleague, Dr. Pamela Popper, has experienced AND's vicious anti-free-speech actions firsthand" (Campbell and Jacobson 2013, 272). From Campbell's perspective Popper has been "bullied by the dietetic industry" and "fought back" (2013, 272). Campbell also writes that he too "received another cold shoulder from medical journals after collaborating with [...] Dr. Alan Goldhamer" (Lisle's co-author) contending that the reason for such dismissals is that medical journals' "income depends on heavy advertising from antihypertensive medicines" leading editors to choose "their wealth over your health" (2013, 235). While such a claim is worthy of discussion, the rhetorical appeals fall back into a bounded discourse community where the conversation is directed via shared assumptions created out of defensive frustrations, somewhat fueled by professional identity anxieties. The narrative becomes one of alternative care practitioners working against the establishment for the benefit of an alienated public.

Regardless of how the medical establishment would contest the credentials of the alternative WFPB diet community, the overall criticism about the corrupting effects that big food and pharma have on the medical establishment is not Campbell's position alone, as other non-vegan researchers and journalist have argued that industry influences of on our dietary practices (see Nestle 2002; Schlosser 2002; Scrinis 2013; Moss 2014). Yet, it will take more than outlier MDs and researchers to promote and mainstream the value of a WFPB lifestyle. Some gradual acceptance among established (mainstream) nutritionists, dietitians, and MDs do exist as many do acknowledge the benefits of a WFPB diet.

A large part of the public is indeed alienated from the scientific and medical establishment as similar lines of persuasion within the vaccine debate have continued to demonstrate, blurring the lines between science, politics, and public perception. Paul A. Offit, an advocate for vaccines and vaccine education, acknowledges this frustration: "when our health is at stake, it's hard to accept that our [scientific] knowledge is incomplete" (2018, 3). Offit points out that pharmaceutical companies, at least when it comes to vaccines, are not cigarette companies (2018, 190). The WFPB community could also push back, in turn, agreeing that indeed, vaccines are not bad for you, but meat, like smoking, is. Also, much like our reliance on oil, a global sustainability impact of non-human animal agriculture appears to be very real (see Oppenlander 2012; Andersen and Kuhn 2015; Hayes and Hayes 2015).

Audience perception is a key concern for any discourse community. As the traditional and alternative health and lifestyle advocates are more inclined to focus on individualized, neoliberal influences (see below), it is the ideological pull of individualism that drives the rhetorical messaging. A cultural ethos based on individualism is hardly revelation and has been a mainstay in how results often undermine intentions, i.e., those appealing to the heart too often end up at a stomach level or what psychiatrist turned marketing guru Clotaire Rapaille would label the "reptilian brain's" response to cultural stimuli (2007, 74–5). Without the larger vegan message of social justice, "plant-based" becomes a term relegated to the status of faddish diets.

Ideological Appeals

Neoliberalism, as a market-driven ethos (see Harvey 2005; Brown 2015), should also be acknowledged as a driving ideological force embracing the post liberation movement backlash. The backlash narrative sees the so-called hippies of the 1960s become yuppies in the free market 1980s and beginning to embrace "the doctrine of individual responsibility" (Ehrenreich 2018, 64). Such a shift, coming out of ideological disenchantment of the 1960 zeitgeist, opens the door to a libertarian ideological frame of reference. The individualistic pursuits of health, self-improvement ethos coming out of the lifestyle movements find vegetarianism at risks of co-optation or reductionism.

Libertarianism best exemplifies a form of neoliberal capitalist ethos, placing a faith in the free market as "an all-purpose pop explanation for virtually everything" (Lehmann 2011, 43). Popper's frustration is channeled through such a libertarian framework as she uses the vocabulary of the right when calling Medicare an "entitlement" (2013, 212) or when advocating that "we cannot afford to continue to grow government" (2013, 208). Why a WFPB lifestyle may be of note in this context is that Popper's advocacy is also promoted as a pushback toward the establishment. Such a libertarian ethos is grounded in an "Objectivist Folk Model of the Self [...] bifurcated into reason and desire" (Johnson 1993, 128–9) where the ability to reason is the core of an individualist drive. Popper's libertarianism is evident when she promotes personal "responsibility for your own health" (2013, 221) while viewing universal health care as form of unlimited government (2013, 222). Such views push the dialogue away from the larger vegan movement.

While careful not to outright criticize capitalism, Campbell, however, finds it "natural for all the elements in a system to do what they can to survive and thrive" (Campbell and Jacobson 2013, 189). Further, the "problem is not inherent to the free market, but rather the result of a market manipulated by its most powerful participants, often through collusion with a government far removed from the people it is supposed to serve" (Campbell and Jacobson 2013, 189). Although such criticisms may be rhetorically savvy, it may also be equally naive of Campbell to believe that we can have markets that are not manipulated by powerful interests. But, again, Campbell is not framing a leftist argument, but one grounded in traditional liberal stance (see Lazere 2005, 361), advocating for market restraints. Some from the left may wish for Campbell to go further, but for his purposes, to channel his frustrations, he remains focused on those corrupting powers within capitalism rather than making a systemic critique. Yet, within the WFPB community, we now see polar political orientations of Popper's overt libertarianism and Campbell's liberalism united under a banner of anti-establishment criticism.

It is how the discussion plays out in the public that separates the WFPB advocates from the rest of the vegan movement. A libertarian ethos can also be seen in Joel Kahn's advocacy, but he comes with more mainstream credentials as a "graduate of the University of Michigan School of Medicine [...] certified in internal medicine, cardiovascular medicine, interventional cardiology, nuclear cardiology, and metabolic medicine" and, as of this writing, "a clinical professor of medicine at Wayne State University School of Medicine" (Kahn 2018, 249). Kahn came into the popular consciousness after his appearance on comedian and mixed martial arts enthusiast Joe Rogan's podcast in 2018 where, for over 3 hours, he debated Paleo diet advocate Chris Kresser (Rogan, Kresser, and Kahn 2018). The appearance on *Rogan* is telling as to Kahn's intended audience: the macho, aggressive masculine ethos typical of Paleo, meat-based advocates, or for vegan males, what Wright labels as "hegan" (2015, 26). With this audience in mind, Kahn also promotes the role that a WFPB diet may play in addressing erectile dysfunction (Kahn 2018, 113–8). Kahn's outreach via *Rogan* attempts to appeal to this individualized and sexualized connection between meat and manhood (see Adams 2015; Wright 2015, 110; Contois 2020), quickly moving from heart disease to erectile dysfunction while also appealing to the masculine power promoted by such an audience.

Kahn's own diet book, *The Plant-Based Solution* (2018) is an overview of WFPB health written for the general public. He argues for a WFPB diet to reverse or hinder various ailments (heart disease, diabetes, high blood pressure, cancer, Alzheimer's). As with Campbell, Kahn also links a WFPB diet to environmental and ethical aspects, but most of the book features case studies of individual clients and recaps of classic and current research. The appeal to an ethos of legitimacy is established in the introduction as Kahn writes "from my place of authority and experience. My oath not to harm that I took as a medical doctor is the principle [readers] can trust" (2018, xix). Kahn does, however, acknowledge animal welfare and the environmental impacts of non-human animal consumption. Like Campbell, Kahn addresses animal agriculture, concentrated animal feeding operations (CAFOs) (2018, 128), animal liberation movements (2018, 131–2), and ethical veganism (2018, 132–6) along with a chapter devoted to the environmental impact that a WFPB diet can have in reducing greenhouse gases (2018, 138).

But in working to reach a public Kahn, in the traditions of Barnard, Campbell, and other WFPB advocates, frames much of his argument with an awareness of the popularity of the low-carb diets. It is this framing that is often the problem that fractures the WFPB diet movement from the larger vegan advocacy. The result of this individualistic ethos is that the mainstream debates between keto and vegan are grounded in similar ideological assumptions that don't address the larger vegan movement (environmentalism and non-human animal welfare). Rather, we can see how, in diet and health centered debates, an ethos of individualism pervades the WFPB health movement, much as with the base level appeal of anti-carbohydrate diet plans from Atkins to Keto to Paleo. Anti-carbohydrate advocates also align with a market oriented, if not outright libertarian, ideological framework, as is predominant in the tone and pathos of Joe Rogan's podcast (Rogan, Kresser, and Kahn 2018). Indeed, Kahn is placing veganism on the same level of lifestyle and diet by arguing with Keto advocates.

Further, Kahn, through his appearance on *Rogan*, is arguing based on the same libertarian and macho ideological framework discussed above, thereby making any appeal to ethical vegetarianism less successful to the *Rogan* audience. Audience acceptance, regardless of any argument, is achieved by "inducing a set of mood and motivations – an ethos [...] a world view" (Geertz 1973, 118). In the case of the Rogan podcast, the world view (or ideological framework) of the audience is off-code for his message. Appealing to the ethos of Rogan's audience won't make for a persuasive WFPB argument, and Kahn falls into the same traps as Barnard describes in the 2000 debates with Atkins adherents (2001, 146). Neoliberal attitudes, which value individual merit over collective action, partly helped make the anti-carb diets so popular (Weisse 2020). Many WFPB advocates are making appeals on a foundation of market-based, entrepreneurial drives (Leary 2018, 82). Such a narrow focus on individual choice leaves out the collective good that comes with veganism. Scientific evidence on its own won't be persuasive without a wider commitment without reinforcing the argument beyond a "healthy lifestyle," mainstream vegan advocacy remains fractured.

Conclusion: The Practice of Veganism

While the *Forks over Knives* documentary brought a wider, mainstream exposure to the WFPB movement, it also leaves us with lingering questions about diet as a prime motivation for practicing veganism. The alternative and medically mainstreamed WFPB advocates have an advantage in that both raise the bar working together widening the scope by expanding the discourse away from individualized health advice. If the individualist arguments are more resonant than arguments and appeals reasoned on animal welfare or environmentalism, then, at least it is it gets folks to the door. What we do once we have the attention of the audience is the next step and beyond the scope of this chapter.

Campbell was an early proponent of replacing "vegan" with "Whole Foods Plant-Based," arguing that he and other cancer researchers "were reluctant to use the words *vegetarian* and *vegan*, or assign a particular value to the ideology that lies behind much vegetarian and vegan practice [...] however noble they may be" (Campbell and Jacobson 2013, 11). He has, however, made inroads to understanding the "interconnected system" rather than a "focus on parts" (Campbell and Jacobson 2013, 288). Specifically, Campbell addresses the effects of that "gross inefficiency of cycling our plant production through animals" (2013, 173). Campbell also acknowledges that he has a "much greater concern today regarding violence done to animals than before" (2013, 170), elaborating a bit on the ethics of concentrated animal feeding operations (CAFOs) and factory farming.

Promoting veganism with a focus on health and lifestyle plays into a larger cultural ethos of individualism and the ideas of personal responsibility when it comes to the control an individual has over healthy practices (i.e., diet). An introduction to veganism we see in *Forks over Knives* gets at the stomach rather than the heart, to paraphrase Sinclair's reaction to the impact of *The Jungle* (Sinclair 1960 [1906], 349). The experts featured in the documentary are defensive and bring a focus on professional frustrations and anxieties about credibility, expertise, and professional acceptance—which all factor into to the bounded discourse the WFPB health community currently finds itself.

If veganism is seen only as a "plant-based lifestyle," then the larger discussion about animal welfare and environment gets lost in what becomes the faddishness of a diet. As a means of introduction, the diet angle is of value to a wider audience, but it is problematized by an ideological foundation that shifts the focus in appealing to a dominant, popular individualistic ethos. Indeed, appealing to such cultural values may do well in recruitment but unless a rhetorical and ideological shift takes place to link up the individualism of the healthy lifestyle movements to a collective ethos that addresses the social and ethical implications of veganism, there will be little advancement in how veganism is accepted in popular, mainstream discourse.

The personalities outlined above all share an entrepreneurial ethos grounded in their outlier status. Promoting health retreats or clinics within a wider argument makes for a more market-based, ideological frameworks, which further distances the diet faction from the more socially progressive vegan movement. Marketing becomes the measure of the message wherein health experts end up promoting diets rather than engaging with larger ethical or environmental issues. Starr reminds us, however, that "Americans are singularly concerned with their individual well-being" (2017, 7) and "independent professionals [...] derive power from the dependent emotional condition of their clients" (2017, 11). A plant-based advocacy tied up with a carbohydrates debate leaves a public with a less than complete picture of veganism. Although the fear of public perception about the preachiness of veganism as an ethical movement may be compounding anxieties, shaping veganism into a diet fad diet alone will not lead to a long-term cultural movement—but the larger issues surrounding veganism might.

Bibliography

Adams, Carol J. 2015. *The sexual politics of meat: a feminist-vegetarian critical theory.* New York, NY: Bloomsbury.

Andersen, Kip, and Keegan Kuhn. 2015. *The sustainability secret: rethinking our diet to transform the world.* San Rafael, CA: Earth Aware.

Barnard, Neal D. 2001. *Turn off the fat genes: the revolutionary guide to taking charge of the genes that control your weight.* New York, NY: Three Rivers Press.

Basov, Suren, ed. 2016. *Social norms, bounded rationality and optimal contracts.* Singapore: Springer.

Brown, Wendy. 2015. *Undoing the demos: neoliberalism's stealth revolution*, edited by Wendy Brown and Michel Feher. Pbk ed. *Near futures.* New York, NY: Zone Books.

Campbell, T. Colin, and Thomas M. Campbell. 2005. *The China study: the most comprehensive study of nutrition ever conducted and the startling implications for diet, weight loss and long-term health.* Dallas, TX: BenBella Books.

Campbell, T. Colin, and Thomas M. Campbell. 2016. *The China study: the most comprehensive study of nutrition ever conducted and the startling implications for diet, weight loss and long-term health.* Revised and expanded ed. Dallas, TX: BenBella Books.

Campbell, T. Colin, and Howard Jacobson. 2013. *Whole: rethinking the science of nutrition.* Dallas, TX: BenBella.

Cialdini, Robert B. 2009. *Influence: science and practice.* 5th ed. Boston, MA: Pearson Education.

Contois, Emily J. H. 2020. *Diners, dudes, and diets: how gender and power collide in food media and culture. Studies in United States culture.* Chapel Hill, NC: University of North Carolina Press.

Cooney, Nick. 2014. *Veganomics: the surprising science on vegetarians, from the breakfast table to the bedroom.* New York, NY: Lantern Books.

Crawford, Robert. 1980. Healthism and the medicalization of everyday life. *International Journal of Health Services* 10 (3): 365–388. https://doi.org/10.2190/3h2h-3xjn-3kay-g9ny.

Ebert, Roger. 2011. "Eat meat and die! Forks over Knives movie review". *Roger Ebert.com* (blog). February 21. rogerebert.com/reviews/forks-over-knives-2011

Ehrenreich, Barbara. 2018. *Natural causes: an epidemic of wellness, the certainty of dying, and killing ourselves to live longer.* New York, NY: Twelve.

Evans, Karin. 1998. Audience and discourse community theory. In *Theorizing composition: a critical sourcebook of theory and scholarship in contemporary composition studies*, edited by Mary Lynch Kennedy, 1–5. Westport, CT: Greenwood Press.

Fulkerson, Lee. 2011. *Forks over knives.* Newtown, PA: Virgil Films.

Gee, James Paul. 2005. *An introduction to discourse analysis: theory and method.* 2nd ed. London: Routledge.

Geertz, Clifford. 1973. *The interpretation of cultures: selected essays.* New York, NY: Basic Books.

Gregory, Jane, and Steve Miller. 1998. *Science in public: communication, culture, and credibility.* Cambridge, MA: Basic Books.

Habermas, Jürgen. 1975. *Legitimation crisis.* Translated by Thomas McCarthy. Boston, MA: Beacon Press. 1973.

Harvey, David. 2005. *A brief history of neoliberalism.* Oxford; New York, NY: Oxford University Press.

Hayes, Denis, and Gail Boyer Hayes. 2015. *Cowed: the hidden impact of 93 million cows on America's health, economy, politics, culture, and environment.* New York, NY: Norton.

Iacobbo, Karen, and Michael Iacobbo. 2004. *Vegetarian America: a history.* Westport, CT: Praeger.

Johnson, Mark. 1993. *Moral imagination: implications of cognitive science for ethics.* Chicago, IL: University of Chicago Press.
Kahn, Joel K. 2018. *The plant-based solution: America's healthy heart doc's plan to power your health.* Boulder, CO: Sounds True.
Kennedy, George A. 1998. *Comparative rhetoric: an historical and cross-cultural introduction.* New York, NY: Oxford University Press.
Kuhn, Thomas S. 1996. *The Structure of Scientific Revolutions.* Third ed. Chicago, IL: University of Chicago Press.
Lazere, Donald. 2005. *Reading and writing for civic literacy: the critical citizen's guide to argumentative rhetoric*, edited by Henry A. Giroux. *Cultural politics & the promise of democracy.* Boulder, CO: Paradigm Publishers.
Leary, John Patrick. 2018. *Keywords: The new language of capitalism* Chicago, IL: Haymarket Books.
Lehmann, Chris. 2011. *Rich people things.* Updated ed. Chicago, IL: Haymarket Books.
Lisle, Douglas J., and Alan Goldhamer. 2003. *The pleasure trap: mastering the hidden force that undermines health & happiness.* Summertown, TN: Healthy Living Publications.
Maag, Chris. 2005. "Who You Callin' a Quack? A tale of strange medicine, cops in ski masks, and Ohio's out-of-control war against natural health." *Cleveland Scene*, May 4, 2005.
Moss, Michael. 2014. *Salt, sugar, fat: how the food giants hooked us.* New York, NY: Random House.
Nestle, Marion. 2002. *Food politics: how the food industry influences nutrition and health.* Berkeley, CA: University of California Press.
Offit, Paul A. 2018. *Bad advice: or why celebrities, politicians, and activists aren't your best source of health information.* New York, NY: Columbia University Press.
Oppenlander, Richard A. 2012. *Comfortably unaware: what we choose to eat is killing us and our planet.* New York, NY: Beaufort Books.
Perelman, Chaïm. 1982. *The realm of rhetoric.* Translated by William Kluback. Notre Dame, IN: University of Notre Dame Press.
Popper, Pamela, and Glen Merzer. 2013. *Food over medicine: the conversation that could save your life.* Dallas, TX: BenBella Books.
Porter, James E. 1992. *Audience and rhetoric: an archaeological composition of the discourse community*, edited by Nancy Sommers. *Prentice Hall Studies in Writing and Culture.* Upper Saddle River, NJ: Prentice Hall.
Rapaille, Clotaire. 2007. *The culture code: an ingenious way to understand why people around the world buy and live as they do*, edited by Lou Aronica. New York, NY: Broadway.
Rogan, Joe, Chris Kresser, and Joel K. Kahn. 2018. "#1175 – Chris Kresser & Dr. Joel Kahn," July 27, 2020, in *Joe Rogan Experience*, 3:47:38, thejoeroganexperience.net/episode-1175-chris-kresser-dr-joel-kahn/
Schlosser, Eric. 2002. *Fast food nation: the dark side of the all-American meal.* New York, NY: Perennial.
Scrinis, Gyorgy. 2013. *Nutritionism: the science and politics of dietary advice. Arts and traditions of the table: perspectives on culinary history.* New York, NY: Columbia University Press.
Simon, Herbert A. 1957. *Models of man: social and rational; mathematical essays on rational human behavior in a social setting.* New York, NY: Wiley.

Sinclair, Upton. 1960 [1906]. *The jungle. Signet classic.* New York, NY: New American Library.

Singer, Peter, and Jim Mason. 2006. *The way we eat: why our food choices matter.* Emmaus, PA: Rodale.

Spencer, Colin. 1995. *The heretic's feast: a history of vegetarianism.* Hanover: University Press of New England.

Starr, Paul. 2017. *The social transformation of American medicine.* Updated ed. New York, NY: Basic Books.

Van Leeuwen, Theo. 2008. *Discourse and practice: new tools for critical discourse analysis. Oxford studies in sociolinguistics.* Oxford, UK: Oxford University Press.

Weber, Max. 1978. *Economy and society: an outline of interpretive sociology.* Translated by Ephraim Fischoff, Hans Gerth, A.M. Henderson, Ferdinand Kolegar, C. Wright Mills, Talcom Parsons, Max Rheinstein, Guenther Roth, Edward Shils and Claus Wittich. 2 vols., edited by Guenther Roth and Claus Wittich. Berkeley, CA: University of California Press.

Weisse, Travis A. 2020. "Through Thick and Thin: Americans' Trust in Dietary Experts *(1945–2005).*" PhD dissertation, University of Wisconsin, Madison (ProQuest 2407596320).

Wilson, Marc Stewart, Ann Weatherall, and Carly Butler. 2004. A rhetorical approach to discussions about health and vegetarianism. *Journal of Health Psychology* 9 (4): 567–581.

Wright, Laura. 2015. *The vegan studies project: food, animals, and gender in the age of terror.* Athens, GA: University of Georgia Press.

5 Agency and Rhetorical Citizenship in Nineteenth Century and Contemporary Vegetarian Discourses in the United Kingdom

Kristin Kondrlik

On September 10, 1855, James Simpson, the president of the British Vegetarian Society delivered a lecture where he decried the state of the common man during an age of technological, scientific, and intellectual progress. He noted that "Man groans in bondage, because ignorance, error, and self-indulgence have filled his flesh with corruption" ("Accrington Vegetarian Association Lecture" 1855, 51).[1] This corruption, Simpson argued, resulted not only from vices such as liquor, lust, or tobacco, but also from diet. Specifically, Simpson asserted that an "erroneous diet" consisting of meat influenced a person's "stomach," "nerves," and "brain," and pervaded their "body," "feelings," and "thoughts" (55). These effects prevented individuals from becoming "aware of the nobility, beauty, height, and capacity of [their] existence" (55). Nineteenth-century vegetarians highlighted vegetarianism's physical and ethical advantages: consuming a plant-based diet strengthened physical and mental health (Morton 1998; Puskar-Pasewicz 2004; Neely 2013). Simpson, however, also argued that meat eaters would be prevented from achieving full happiness: "[T]here are moral facts and moral being, which can never be revealed, and degrees of moral happiness that cannot be enjoyed, till all the laws of harmonious relationships are fully obeyed" (55). For Simpson, then, eating a vegetarian diet was not only an issue of animal rights or individual morality. Rather, Simpson and other members of the nineteenth-century British Vegetarian Society wrote that the moral, intellectual, and physical effects of eating meat could prevent an individual from participating in a "harmonious" relationship with animals and the environment.

Many contemporary vegetarians and vegans[2] retain Victorian vegetarians' idea of the connection between an individual's diet and their ability to contribute to society. Analyzing contemporary discourse, scholars note that today's vegetarians and vegans frequently adopt a meat-free diet for individual health or ethical concerns (Jorgensen 2015). Increasingly, however, advocates argue for vegetarianism and veganism on the explicit basis of "citizenship." Micheletti and Stolle (2012), for example, note that activists frequently invoke the concept of "sustainable citizenship," which recognizes the responsibilities of the individual to local and global communities on issues of environmental, economic, and social justice.

DOI: 10.4324/9781003039013-5

To map the evolving relationship between vegetarianism and citizenship, this chapter will analyze the writing of the British Vegetarian Society in the nineteenth century and today. To this end, this chapter will examine the Victorian *Vegetarian Messenger* and the web publications of the contemporary Vegetarian Society of the United Kingdom. These historical and contemporary discourses are connected to what Kock and Villadsen (2014) have termed "rhetorical citizenship": the idea that utterances performed by citizens and non-citizens in a particular society move individuals to act and enact the formation and maintenance of public life.

Over time, the role of citizenship, as represented in the Vegetarian Society's official publications, has shifted. Both historical and contemporary discourses share concerns about diet and citizenship but differ in their articulations of the contributions of non-vegetarians. Nineteenth-century sources focused on the moral, physical, and mental transformations caused by eating meat and argued that consuming animal products corrupted non-vegetarians. This corruption inhibited an individual's agency and rational decision-making, destroying his ability to serve community interests. Through these arguments, the nineteenth-century movement asserted that those who enacted citizenship by refraining from meat could improve the state of society, while those excluded from vegetarian citizenship could not. The Society's contemporary writings, however, center more squarely on the embodied and discursive responsibilities incumbent upon "citizens" and "non-citizens" of the vegetarian community. The contemporary Society's website adopts the attitude that non-vegetarians can make public arguments and contribute to the well-being of civic society by limiting their consumption of meat. Therefore, while the constraints upon the definition of a "vegetarian citizen" have remained largely the same, with changes to the scientific evidence it relies on for its arguments, modern vegetarianism more clearly recognizes the contributions that non-citizens can make both to the vegetarian community and more broadly to global society.

Rhetorical Citizenship

The question of who has a right to speak (and who does not have the right to speak) and how "citizens" can best exercise that right has long been a concern of rhetorical studies. We live in times, however, of unstable perspectives on who can and cannot be a citizen. Addressing this instability, Robert Asen outlines a "discourse theory of citizenship." Rather than seeing citizenship as a fixed identity, Asen argues that scholars should move to re-frame "citizenship as a fluid, multimodal, and quotidian process" (2004, 203). Such a shift, Asen argues, better reflects the reality that "[c]itizenship engagement cannot be distilled to a set of rights, condition of membership, or allegiance to a cultural tradition" (2004, 203). Rather, Asen writes, "[P]eople enact citizenship through their own agency. This means, too, that citizenship may be enacted by non-citizens" (2004, 204). Other scholars, including Karma R. Chávez, have outlined citizenship practices unconstrained by legal and political boundaries: civic, community, and activist practices, and sexual, social, consumer, and cosmopolitan citizenship (2015, 164–5). Participation in a community can still be denied individuals who claim legal or cultural citizenship.

Christian Kock and Lisa S. Villadsen echo Asen's movement away from nation-bound citizenship. They have argued that scholars looking to understand "rhetorical citizenship" should analyze exercises of and limitations placed on rhetorical agency: "citizens' possibilities for gaining access to and influencing civic life through symbolic action" (2014, 10). The authors affirm that "legal rights, privileges and material conditions are not the only constituents of citizenship"; rather, they note that "discourse that takes place between citizens is arguably more basic to what it means to be a citizen" (2014, 11). In this view, who possesses the power to speak or perform symbolic actions as a "citizen" has greater force than a legal document affirming that individual's citizenship. For example, an immigrant to the United States may win the legal right to call herself a citizen, but her ability to participate in public discourse may be undermined through acts of silencing or intimidation. Thus, her ability to affect change in society through utterances or symbolic actions has been reduced. Citizenship itself, then, is rhetorically contingent. Analyzing this contingency, Kock and Villadsen argue, can lay bare the connections between specific discursive or symbolic actions of citizens and the "maintenance and development of civic life" (2014, 13).

These citizen discourses, however, are not limited to written and spoken utterances. Rhetorical studies scholarship takes seriously the role of embodied discourse, or the use of the body to participate in public conversations. From voting to dancing, from running to protesting, scholars have argued that "speakers" make public arguments with their bodies (DeLuca 1999; Ray 2007; LeMesurier 2014; Samek 2017). Considering such scholarship, it is not a leap to recognize how eating or not eating meat can be used as an exercise of one's right to "speak" as a citizen. Eating an animal can limit or contribute to an individual's recognition as a citizen of a given community. This was especially true in the early years of modern vegetarianism, which framed abstaining from meat as key to citizenship in a utopian vegetarian community and to reaching humankind's full potential.

Eating the Way to an Enlightened Citizenry: Vegetarianism in Victorian Britain

As Margaret Puskar-Pasewicz has noted, vegetarianism has been practiced worldwide since ancient Eastern religions and Stoic philosophers (2010, viii). Modern vegetarianism in the United States and Britain, however, is rooted in eighteenth-century humanism (Spencer 1995, 253–4; Gregory 2007, 2; Simons 2010, 231). The vegetarian movement formalized as the nineteenth century progressed, and the first western vegetarian organization was formed in September 1847 as the British Vegetarian Society. The Society gained popularity in its first years but remained a relatively small organization throughout the nineteenth century; membership in the Society increased from 478 in 1850 to 889 in 1853 (Spencer 1995, 266). Though the Society continued to grow, the public still viewed it and its members as radical. The Society's founders and members pushed against societal norms that framed Britain as a nation of carnivores, and its public reputation followed in this line. As James Gregory has noted, Victorian Britons largely regarded vegetarianism as one of many passing health fads lead by quacks and fanatics (2007, 119–20). The current Vegetarian Society of the United Kingdom and the Vegan Society are direct descendants of this organization. However, as we will see, the current iteration of the Vegetarian Society adopts a much less radical approach than its ancestor.

The British Vegetarian Society documented its early years in periodical publications. The first publication associated with the Society was the *Vegetarian Advocate*. Though the *Vegetarian Advocate* initially positioned itself as the Society's vanguard publication, the Society's president James Simpson soon established the *Vegetarian Messenger* (Spencer 1995, 262). In its first years, Simpson and the Society's secretary oversaw the publication. The Society's modest membership, however, could not sustain two periodicals. Therefore, in March 1851, the *Vegetarian Advocate* published its final issue. For the next decade, until James Simpson's death in 1859, the *Vegetarian Messenger* served as the primary voice of the burgeoning Society. Examining the first decade of publication reveals some of the major ideas that grounded the movement as it solidified in the mid-nineteenth century.

Unsurprisingly, the *Vegetarian Messenger* reveals Victorian vegetarianism's primary goal as encouraging individuals to adopt a vegetarian diet, but the first issue of the *Vegetarian Messenger* listed a variety of objectives. At its heart, Simpson placed the idea that humans, as "physical, intellectual, and moral being[s], can become most completely developed, in all [their] faculties, by subsisting upon the direct productions of the Vegetable kingdom" ("The Vegetarian Movement. Its Claims upon Public Attention" 1851, 2). Humankind, then, would be transformed into the best version of itself—intellectually, physically, and morally—by refraining from the consumption of meat. Simpson repeated this call for humanity's betterment through diet frequently throughout the journal's first years.[3] The argument that a vegetarian diet promoted better physical, mental, and moral health will not be unfamiliar to modern readers. However, Victorian vegetarians' arguments also looked beyond their own health and happiness and toward the creation of an idealized society with a more enlightened citizenry—achieved by the universalization of the vegetarian diet.

Victorian vegetarians, notably, flirted with utopian ideals popular during the nineteenth century. Kathryn R. Falvo has acknowledged the connection between nineteenth-century utopianism and vegetarianism through the American example of Fruitlands: an agrarian commune founded by

Bronson Alcott and Charles Lane in the 1840s. Fruitlands, Falvo notes, aimed to "liberate" the land and radically alter the relationship between humans, animals, and nature (2019, 38). Alcott and Lane focused on reforming society through self-improvement, facilitated by individual behavioral reforms. Falvo writes that decisions regarding diet "were at the very center of forming new humans and hence a new world" (2019, 39). Alcott especially was invested in helping individuals reach their full potential, and Falvo notes, "Creating utopia, for [Alcott], was not about creating a perfect world. It was about creating perfect humans, who were the backbone of that perfect world" (2019, 39). Instead, Falvo argues, vegetarian practices were "not about animals at all. [They were] about re-creating humans" (2019, 39).

In much the same way as Fruitlands focused on individual reform, the members of the British Vegetarian Society centered their discussion on how vegetarian practices might aid in creating an ideal civilization. In the *Vegetarian Messenger*'s January 1851 issue, Simpson, like Alcott and Lane, envisioned a society that more closely approximated humankind's full potential and, through the "civilizing spirit of [the] age[,]" progress toward "a truly enlightened and merciful civilization" ("The Vegetarian Movement. Its Claims upon Public Attention" 1851, 4). Simpson wrote extensively about the nature of this idealized society. In the first issue of the journal, he noted that one of the Society's major tasks was to "prov[e] the many advantages of a Physical, Intellectual, and Moral Character, resulting from *Vegetarian Habits of Diet*; and thus to secure, through the Association, Example, and Efforts of its Members, the adoption of a Principle, which will tend essentially to *True Civilization*, to *Universal Brotherhood*, and to the *Increase of Human Happiness*, generally" ("The Vegetarian Society" 1849, n.p.). Like Alcott and Lane, Simpson argued that reforming individual behaviors, particularly through diet, would create a more harmonious society.

The establishment of this "True Civilization" relied on the formation of idealized citizens. In January 1850, Simpson compared this utopia to the interconnectedness of a tree: "… every leaf, fibre, and vein has its service to perform to complete the harmony of its nature, should not every thought, word, and action of our lives, likewise serve to preserve the harmony of our creation?" ("The Vegetarian Principle. Lecture II: Economy; Intellectual and Moral Advantages" 1850, 27). Here, Simpson laid out a precursor to Micheletti and Stolle's notion of sustainable citizenship, which charges individuals to "do all they can to help improve social justice and safeguard nature to make the world a better place to live" (2012, 89). Specifically, Micheletti and Stolle argue that sustainable citizenship expands the responsibilities of individuals and businesses and note that it "views both private and economic life as important centers in the making and performance of citizenship" (2012, 90). Members of the historical Vegetarian Society argued that individual vegetarians contributed to the well-being of society by making embodied discursive choices to avoid meat. However, if we draw on Kock and Villadsen and analyze how citizens and non-citizens contribute to or maintain civic life through such discursive actions, the picture of citizenship in Victorian vegetarianism darkens. The very act of meat-eating, Victorian vegetarians argued, could rob an individual of their ability to contribute to "True Civilization." Indeed, Victorian vegetarians often went so far as to argue that eating meat robbed individuals of their humanity. For these writers, eating meat sullied eaters and their contributions to civic life: first, meat eaters became morally corrupt by participating in the Victorian meat economy; and second, meat eaters damaged their physical and intellectual faculties. Thus, only full citizens of the vegetarian community—those abstaining completely from meat—could contribute to the betterment of civic life.

Meat-Eating as a Moral Transgression with Long-Term Effects on Citizenship

Anticipating Micheletti and Stolle's emphasis on the obligations on consumers and businesses, the *Vegetarian Messenger* condemned Victorian Britain's meat economy as hindering societal harmony. The violence at the heart of this industry, writers argued, would harm society's interests by encouraging immorality. As editor, Simpson decried animal slaughter as moral transgression. In the first issue of the journal, Simpson wrote that animal slaughter represented "unnatural and cruel conduct of man toward those he was designed to protect" ("Introduction" 1849, 1). Similarly, in June 1855, Simpson commented that the sanctity of all life—animal life included—"belong[s] to the very genius of Vegetarianism" ("The Preying upon Animals the Trainer for War" 1855, 46). He argued that a vegetarian lifestyle "[guarantees] society against many of the gross and glaring evils and anomalies which now produce its greatest sufferings and misery" ("The Preying upon Animals the Trainer for War" 1855, 46).

In addition to animal slaughter, Simpson also decried animal husbandry. In January 1850, he noted that livestock farming was costly and "pollute[d] the air by poisoning our rivers and streams, the banks of which, instead of being the most pleasing resorts for our population, [were], in many instances, rendered highly offensive and unhealthy" ("The Vegetarian Principle. Lecture II: Economy; Intellectual and Moral Advantages" 1850, 30). He also outlined the ecological consequences of livestock farming, including famine, poverty, and waste. He lamented that while "thousands and tens of thousands of our fellow-creatures are in a state of privation and want," animals were fattened for slaughter "with the produce of our soil" ("The Vegetarian Principle. Lecture II: Economy; Intellectual and Moral Advantages" 1850, 30). Simpson condemned Britain's meat industry as immoral not only for its cruelty to animals and the environment but also for the economic damage it imposed on less privileged human beings.

The stain of moral degradation applied not only to the meat industry but also to consumers. Simpson lambasted popular "live" markets—where many Londoners purchased their food—as dangerous and depraved. In

January 1850, Simpson declared, "If we wish to see depravity more depraved, and barbarism in the midst of civilization; if we wish to see cruelty and crime carried on to the most revolting depths; where can this wish be so well gratified, as in our Smithfields, our Leadenhalls, and our Billingsgates?" ("The Vegetarian Principle. Lecture II: Economy; Intellectual and Moral Advantages" 1850, 35). Simpson indicted these markets—centers of animal slaughter, of crime and violence, and in the case of Smithfield, of historical executions—as hubs of moral corruption and barbarity.[4] These degradations extended beyond the industry and to consumers. Simpson argued that "the practices pursued in these localities have a tendency to lower the moral standard, and to exalt the brutal and degrading part of human nature[,]" as well as to "[destroy] all the noblest sympathies of human nature" ("The Vegetarian Principle. Lecture II: Economy; Intellectual and Moral Advantages" 1850, 35). Clearly, for Simpson, the moral harm perpetuated through a carnivorous diet was not simply humans ignoring pollution or animal suffering; rather, he worried about the lingering damage that proximity to this industry would have on a consumer's ability to achieve the ideals of citizenship prized by the vegetarian movement.

Simpson argued that eating, buying, raising, hunting, or butchering meat had long-term implications not only for morality but also for agency and rationality. Being party to the slaughter of animals, Simpson argued in June 1850, would "blunt" people's feelings so that "what is an act of cruelty to those unaccustomed to such scenes, appears to them as a matter of ordinary business" ("Fruits and Farinacea the Proper Food of Man" 1850b, 90). In January 1851, he similarly asserted, "If the mind become, by habit, accustomed to a disregard of moral convictions, the conscience becomes seared, and does not act so powerfully in guarding the mind from its grosser tendencies; and thus, that which would become the greatest blessing, may be turned into a dreaded curse!" ("The Vegetarian Movement. Its Claims upon Public Attention" 1851, 4). Eating meat, much like slaughtering an animal or hunting for sport, Simpson warned, would impede an individual's ability to make moral choices. Such was the case even if they did not raise, farm, or slaughter animals. Simpson argued that by becoming inured to the industry's violence, meat eaters placed constraints on their morality and agency, foreclosing them from full citizenship in the harmonious civilization that Victorian vegetarians sought.

Meat-Eating, Embodiment, and Citizenship

Concerns about agency and morality also extended to the physical act of consuming meat, and Victorian vegetarians related their arguments to medical and scientific discourse. As Haejoo Kim has noted, Victorian vegetarians argued that eating meat caused physiological changes: one's body would be "morphologically readjusted by one's diet" (2021, 527). Their diet, writers like Simpson argued, put meat eaters at higher risk for disease, and writers characterized meat as a poison or toxin. The March 1855 issue, for example, noted that "Flesh-eating renders the body much more liable to sickness. How can persons be healthy who are every day, and at every meal, swallowing the seeds of disease?" ("Flesh-eating and Its Concomitants" 1855, 21). The *Vegetarian Messenger* positioned even the best quality animal meat as injurious, arguing that meat-eating produced heart palpitations, inflammation, dyspepsia, spasms, fatigue, and ill temper.[5] Simpson collected a variety of evidence—from vegetarians, newspapers, and doctors—that suggested that a vegetarian diet decreased the risk of conditions such as tuberculosis, asthma, scurvy, cholera, and depression, among others.[6] Further, in a period before close food regulation in Britain, vegetarians warned about food contamination. This was not an unfounded fear, given the prevalence of parasitic trichinella in pork imported from the United States and Germany.[7] Britain's meat was not immune from adulteration, especially through the continuing illicit sale of "slink" meat—or meat unfit for human consumption.[8] Simpson and many correspondents noted that a vegetable diet would lessen the risk of consuming adulterated meat and offered numerous anecdotal examples of the outstanding health of vegetarians.[9]

More strikingly, Victorian vegetarians also argued that eating meat could harm carnivores' mental faculties and morals, leading them into a degraded state. For Victorians in general, the functioning of the body and the functioning of the mind were intimately connected. As Jenny Bourne Taylor and Sally Shuttleworth have noted, perceptions of physical and mental boundaries shifted in the nineteenth century; much of the materialist science of the era "rejected the dualistic division between mind and body" (1998, xiv). The *Vegetarian Messenger* featured many anecdotal accounts that reiterated this belief in a mind-body connection. In March 1850, an unnamed writer detailed the story of a Scotsman who strayed from vegetarianism. The Scotsman had grown up vital, intelligent, and hearty, eating a primarily vegetarian diet. As he aged, however, he increased his meat consumption. The Scotsman detailed the mental changes that resulted from a meat-based diet: "At first, I began to feel dull and stupid, especially after eating; my memory became so bad, that the attempt to call anything to my recollection was exceedingly painful; constant head-ache set in, accompanied with a determination of blood to the head – my face flushing if I stooped to the ground" ("Experimental Philosophy" 1850, 59). He noted that meat-eating made him "miserable[,]" "low spirited[,]" and "under a state of constant depression" ("Experimental Philosophy" 1850, 59). On medical advice, he cut back on meat and found that his vitality and mental clarity returned. He asserted, "*I felt as if I could leap over the house tops, and shake hands with every one I met*" ("Experimental Philosophy" 1850, 59, emphasis in original). Eating meat hindered the Scotsman's physical and mental functioning. Later, in June 1850, Simpson would reiterate the vegetarian diet's mental benefits: "My conviction is, that all who judge reasonably and practically of the Vegetarian system, will not merely find it imparts a more 'pleasing consciousness of existence,' but will find it characterised by greater endurance in both mental and physical occupations" ("Experience of Vegetarians" 1850, 6). To reap these benefits, an individual must avoid meat.

If refraining from meat imparted greater mental and physical fortitude, consuming meat prevented humans from reaching their full potential—intellectually, morally, and spiritually. In December 1849, Simpson asserted that meat-eating would make the intellect "beclouded," the moral senses "blunted," and "our real and substantial enjoyment very much abridged" ("The Vegetarian Principle. Lecture I: Introductory, Physical Advantages" 1849, 22). Simpson went so far as to argue that eating meat could "impair" the nervous system, writing that the "victim" of this "dreadful disorder" suffers from "the most awful despondency" and that they may find life "tedious and wearisome[,]" become apathetic, and consider or even commit suicide ("The Vegetarian Principle. Lecture I: Introductory, Physical Advantages" 1849, 22).

The "Accrington Vegetarian Association Lecture" demonstrates some of the additional intellectual and moral limitations purportedly resulting from a carnivorous diet. Simpson wrote that a person "cannot become aware of the nobility, beauty, height, and capacity of his existence while an erroneous diet influences his stomach, his nerves, and his brain, and pervades his body, his feelings, and his thoughts" ("Accrington Vegetarian Association Lecture" 1855, 55). Further, in this same lecture, Simpson argues that humans must "obey" the "laws of harmonious relationships" in order to achieve their full intellectual and moral potential ("Accrington Vegetarian Association Lecture" 1855, 55). An individual who consumed meat could *never* be a full citizen of Victorian vegetarians' idealized community not only because meat-eating imposed obstacles to physical and intellectual well-being, but also because this impaired functioning prevented a moral and harmonious relationship with other living things.

Beyond excluding carnivores from their utopian idyll, Victorian vegetarians argued that eating meat transformed a person into something less than human: placing them in a "primitive," animalistic state. As Kim has argued, "In Victorian vegetarian rhetoric, the individual is often engaged in a moral struggle with a society that drags them down to an 'unnatural' or 'primitive' state of meat consumption"—one that causes physical and moral "degeneration" and hampers their individual liberty (2021, 527). A belief in the dangers of "stimulating" food and drink was one Victorian vegetarians shared with temperance advocates; both argued that "stimulating" substances—alcohol, tobacco, and "exciting" food (spices, mustard, meat, tea)—could radically alter behavior. In this vein, Victorian vegetarians argued that eating meat would excite "animal impulses." In January 1851, Simpson argued that following vegetarian principles would bring "the animal propensities of man into subjection to the moral and intellectual faculties" ("The Vegetarian Movement. Its Claims upon Public Attention" 1851, 3). As evidence of this point, Simpson gestured to scientific writing on the behavioral differences between different animals, noting that "carnivorous races of animals are naturally of impetuous or ferocious dispositions; whilst the herbivorous are mild, and docile" ("The Vegetarian Movement. Its Claims upon Public Attention" 1851, 3). Such, Victorian vegetarians argued, was also the case for humans.

To illustrate this point, the *Vegetarian Messenger* described meat consumption as making humans violent. In January 1850, Simpson noted that eating meat was often associated with "the savage, the ferocious, and cruel dispositions" ("The Vegetarian Principle. Lecture II: Economy; Intellectual and Moral Advantages" 1850, 33). Simpson further asserted that "living on prey" would make humans "carnivorous, ferocious, or unclean in our dispositions, practices, and desires" ("The Vegetarian Principle. Lecture II: Economy; Intellectual and Moral Advantages" 1850, 34). The periodical also featured selections from other writers as evidence of this phenomenon. In March 1850, the periodical printed an excerpt from Jean-Jacques Rousseau's posthumous novel *Emilius and Sophia* (1783). The passage argued that "those people who are great eaters of meat, are, in general, more ferocious and cruel than other men. This observation holds good of all times and all places: the English barbarity is well known, whereas, the Gaures (who abstain from flesh), are, on the contrary, the meekest creatures in the world" ("Ferocity and F[l]esh-eating" 1850, 20). Rousseau noted that the gentle and meek residents of Gaure, located in the Midi-Pyrenees region of France, did not eat meat. However, Rousseau characterized the English, who consumed a diet of meat and vegetables, as "barbarous," and his use of the term "ferocious" ties meat eaters to a "feral" state and thus to animality.[10]

The periodical further demonstrated this correlation between meat-eating and barbarity by drawing on cutting-edge cultural and scientific discourses. Kim highlights the role of evolutionary science and anthropology in the writing of Victorian vegetarians, arguing that they drew on comparative anthropology, which applied evolutionary theories "to the study of human civilization" (2021, 524). The periodical provides numerous anthropological accounts of other societies to make the case for vegetarianism. For example, in July 1854, the *Vegetarian Messenger* printed excerpts of a travel diary by German novelist Frederick Gerstäcker. The account details Gerstäcker's encounter with Indigenous inhabitants of Argentina. The most striking passage is the following:

> [T]he inhabitants of this country, living on flesh and nothing but flesh, continually engaged in slaughtering, and everywhere surrounded with blood and decomposition, have themselves grown savage and bloodthirsty; and that too often they give proof that they hold the life of a man in little higher estimation than that of a bullock or a horse. Living exclusively on animal food *must necessarily tend to make man ferocious*; and the hand once used to the knife, becomes too familiar with it.
>
> ("The Slaughter Houses of Buenos Ayres" 1854, 65)

Eating animal flesh, Gerstäcker asserted, made these Indigenous people "ferocious." Those consuming only meat lost their respect not only for animal life, but also for the lives of other humans. The *Vegetarian Messenger*, thus, used this account to characterize meat eaters as unable to live in the harmony between humans, animals, and the environment that vegetarians craved, and as displaced from humanity's "natural" place in this hierarchy.

A descent into "ferocity" and away from "civilized" behavior was not limited to anthropological accounts of "savage" nations, however. In November 1854, a Mr. White noted that eating animals put the eater into a "febrile" state, while the mind had "its tone, its elasticity, and its freshness weakened through sympathy with the body" ("Diet in Relation to the Mental and Moral Powers" 1854, 103). Eating meat, for White, strained the body and mind. White also noted that meat-eating English possessed a "choleric" disposition ("Diet in Relation to the Mental and Moral Powers" 1854, 105). Here, White invoked outdated humoral science, which argued that bodily imbalances led to physical and emotional distress. White connected eating meat to choler, or yellow bile, associated with irritability, aggression, and ill temper. He asserted that eating meat "induce[d] a ferocity of temper, unknown to men whose food is chiefly drawn from the vegetable kingdom" ("Diet in Relation to the Mental and Moral Powers" 1854, 105). Thus, White argued that eating meat "is a practice incompatible with the highest mental enjoyment, and consequently with the highest happiness and delight of which man is capable" ("Diet in Relation to the Mental and Moral Powers" 1854, 103). Though the English were not typically on a full-meat diet, according to White, by consuming *any* meat, the English too would become "ferocious" and thus, limit their potential.

By using emerging science to assert that eating meat made one "savage," the *Vegetarian Messenger* called on a specific colonialist and racist public sentiment circulating during the nineteenth century. Though idea of a distinction between non-white, non-European "savages" and "civilized humans" had its roots much earlier, the nineteenth century saw philosophers and scientists crystalize theories about the relationship between these two figures. As Robert J.C. Young has argued, John Stuart Mill was one of the first thinkers to place them into a proto-evolutionary hierarchy: for Mill, the savage represented a less advanced, inferior version of the "civilized" man (1995, 35).[11] Carolyn A. Conley has argued that, for Victorians, "civilized" behavior "was what distinguished Englishmen from others. The worst criminals were described as acting like foreign barbarians" (2005, 777). By declaring that an individual was made a savage by meat-eating, Simpson and other vegetarian advocates associated the meat eater with nineteenth-century notions of criminality, inferiority, and violence. Even in supposedly "civilized" cultures such as Britain, the meat eater's behavior disqualified him from full citizenship in "civilized" society and placed him closer to unreasoning animals.

Simpson and other writers linked meat-eating not only with "different orders" of humans but also with a lack of agency. A correspondent to the *Vegetarian Messenger* argued that the "perverted appetite" and "unnatural passion for slaughter" placed a person in "fetters," a "yoke of bondage," and "thralldom" ("Take Up Your Freedom" 1850, 17). Similarly, Simpson wrote that eating meat developed the "gross and depraved part of human nature" ("The Vegetarian Principle. Lecture I: Introductory, Physical Advantages" 1849, 23). He asserted that killing animals and eating meat "degrade[s] and enslave[s] mankind, who thus become easy prey of passion, avarice, and crime" ("The Vegetarian Principle. Lecture I: Introductory, Physical Advantages" 1849, 23). Likewise, in May 1852, Simpson noted that such an appetite for meat would "perpetually subjugate individuals [...] to conditions and states of mind, which, if properly looked into, cannot but be considered as chargeable to the want of conscientious consideration, even in those who profess to regulate their conduct in accordance with the highest moral standard of truth" ("Social and Moral Responsibility" 1852, 35). In other words, individuals became subject to their cravings for animal meat, rather than active subjects themselves. These appetites inclined individuals to satisfy those appetites, rather than to act or speak rationally. In February 1850, Simpson wrote: "It is by departure from [God's] laws that man thus becomes a slave to circumstances. Man is created free, but his choice of evil renders him a slave" ("Fruits and Farinacea the Proper Food of Man" 1850a, 42).[12] Unlike the chattel slavery practiced in the United States at the time, which many vegetarians vehemently opposed, Simpson argued that enslavement to the senses was a state of bondage from which individuals could escape if they chose a vegetarian diet.

"Choosing" to be a slave to one's appetites had serious ramifications for the meat eater's ability to participate in society; the ideal citizen should not allow their intellect and judgment to become overwhelmed by the pleasures of the senses. Simpson declared that being "ruled" by animal nature severed the individual from the best parts of humanity: "Can, then, a practice be conformable to reason and truth which stifles the best feelings of the human heart?" ("Accrington Vegetarian Association Lecture" 1855, 55). If overwhelmed by appetites, Simpson asserted, humans could not contribute to society, because they willingly chose a path of violence inimical to the more reasoned and self-controlled society that Victorian vegetarians sought.

Eating meat, for Simpson, made humans vicious and rash, where they were naturally gentle and mild. Adopting a vegetarian lifestyle, conversely, would help people not only avoid moral, bodily, and mental corruption, but also allow them to be citizens of the vegetarian community and participate in its utopian vision. In his lecture to the Accrington Vegetarian Association, Simpson asserted, "The true man will endeavour to remove every obstruction, that all the benign influences of truth may grow and fructify within him. We may then enjoy life's greatest blessings – a healthy body, a sound understanding, a benevolent heart, and a truth-seeking spirit, ever progressive in all good" ("Accrington Vegetarian Association Lecture" 1855, 55). Britons, according to the *Vegetarian Messenger*, could move beyond the obstruction of their "choleric" temperament and animal appetites and toward full citizenship in a more peaceful and harmonious society. All they had to do to was drop meat from their diet.

Contemporary Vegetarian Discourse: Vegetarian Diets and the Obligations of Citizenship

Public attitudes toward vegetarianism have evolved since the nineteenth century. A 2018 United Kingdom Food Standards Agency report found that over one in ten survey respondents reported adopting a vegetarian, partly vegetarian, or vegan diet (NatSen 2019, 6). Despite this expansion, the core principles of vegetarianism remain largely unchanged. In a 2019 annual report, the Vegetarian Society of the United Kingdom listed its main mission as "to inspire, educate and enable people to be vegetarian" (Vegetarian Society of the United Kingdom 2019, 2). Their vision is "A world where everyone is vegetarian" (Vegetarian Society of the United Kingdom 2019, 2). Strikingly, the report echoes the benefits of a vegetarian diet in the *Vegetarian Messenger*: healthier and happier individuals; a greater sense of community with animals; reduced economic inequality; and more sustainable foodways (Vegetarian Society of the United Kingdom 2019, 2). The contemporary Society focuses more on the connection between diet and issues such as economic equality and environmental destruction, particularly as the realities of climate change become increasingly evident. Though it focuses less on the creation of a utopia in its marketing, the contemporary Society, much like its ancestor, seeks the creation of a more moral, healthier, and happier society through the embodied discursive acts of vegetarianism.

Examining the views of contemporary British vegetarians, however, is more complicated than examining those of Victorian vegetarians. Contemporary communication about social and cultural issues is diffuse and varied; modern media spaces are subject to less gatekeeping, and more individuals have access to technologies that facilitate participation in a print or digital public square. To trace a coherent history, therefore, the remainder of this chapter will focus on the corporate website of the Vegetarian Society of the United Kingdom (https://vegsoc.org/). This website, like the *Vegetarian Messenger*, provides various resources: it features videos and articles about the benefits of vegetarianism, recipes, and articles about social and political issues. Members of the historical and contemporary Society agree that "citizens" of the vegetarian community contribute to the well-being of society by refraining from the consumption of meat. However, in their discussions of non-citizens, historical and contemporary vegetarians diverge. In a reflection of Asen's notion of rhetorical citizenship, contemporary vegetarians recognize that non-citizens can affect positive changes to civic life through their dietary choices—embodied discursive practices. Throughout their materials, the contemporary Society emphasizes that all people are a part of a global community that grants rights and imposes obligations, whether or not they are "full citizens" of the vegetarian community.

The historical and contemporary Societies share similarities, especially in expounding on the advantages of vegetarianism for the eater. Like the historical Society, the contemporary Society emphasizes the physical benefits of a plant-based diet. Their website asserts, "Vegetarian food can be really good for your health" ("Why Go Veggie?" n.d.). It assures readers that a vegetarian or vegan diet can include the same nutrients included in meat and animal byproducts and that a vegetarian diet is "naturally high in fibre and low in saturated fat" ("Health and Nutrition" n.d.). It also features several articles that highlight recipes, tips for balancing nutrients, and testimonials from celebrities, chefs, and social media influencers. This is not unlike the *Vegetarian Messenger*, which collected anecdotes about health improvements, testimonials from reformed eaters, and vegetarian recipes.

To promote vegetarianism's health benefits, the Vegetarian Society's website provides an animated cartoon about a man who becomes vegetarian and improves his health. A video titled "Alan wishes he loved his waistline" details the story of Alan, a middle-aged man who, due to the requirements of adult life, has let his attention to his physical health slip. The narrator describes Alan as feeling "under the weather" and notes that Alan's "step didn't feel quite as springy as it had been before the big 4-0" ("Alan Wishes He Loved His Waistline" 2014). Alan cooks a vegetarian meal, and the video ends with Alan, his wife, and their friends toasting Alan's culinary success. The video largely focuses on the physical benefits of a vegetarian diet, noting studies about reduced risks of cancer, diabetes, and heart disease, as well as the potential for restoring the "spring" in Alan's step. The video calls to mind the *Vegetarian Messenger*'s tale of the Scotsman who abandons and then re-adopts a vegetarian diet, though the video stops after Alan cooks his first vegetarian meal, rather than tracing long-term effects.

Although they both emphasize vegetarianism's benefits for health, the attitudes toward citizenship in historical and contemporary vegetarianism are distinct, especially in how they discuss the connections between vegetarianism's embodied effects and civic participation. Notably absent from the contemporary Society's website is the early Society's focus on meat-eating and a loss of intellectual ability or reason. The video, articles, and website do not make the argument, as the *Vegetarian Messenger* does, that eating meat will make an individual violent, muddled in their thinking, depressed, controlled entirely by "animal appetites," or unable to contribute to society. In contrast to its early years, similarly, the contemporary Society does not conflate meat-eating with the loss of humanity. Notably, the site drops the Victorian Society's engagement with outdated scientific arguments in cultural and evolutionary anthropology. In both cases, the Society offers insights that are in step with emerging medical and social science. However, the Victorian Society's focus on meat-eating as transforming an individual into a "lesser order" of human is clearly bound up in the explicitly imperialist and racist overtones of Victorian medical and social science. Conversely, the contemporary website eschews this dismissal of meat eaters. The website discusses individual health behaviors and health benefits in ways that do not impugn the humanity of the meat eater, even if they cannot achieve full citizenship in the community of "vegetarians." The contemporary Society, instead, takes a more inclusive approach when focusing on the physical and mental effects of eating meat: it argues that reducing the consumption of meat can benefit an individual's physical well-being, as well as their community.

Rather than suggesting that meat eaters necessarily exclude themselves from a harmonious community of humans, animals, and the environment, the contemporary Society instead leans on the obligations of every individual to the global community, rather than on securing them as a "citizen" of the community of vegetarians. Like the Victorian Society, the contemporary Society's website emphasizes the connections between meat consumption and animal cruelty. The website details the harm inflicted on animals slaughtered for food in the United Kingdom. It notes that, although most animals are stunned before death, "still many animals die in fear and pain. For many vegetarians, no form of slaughter can ever be considered humane" ("Going Veggie for Animals" n.d.). It takes pains to note that pigs, "intelligent and inquisitive animals[,]" are crowded into small living spaces and prevented from performing natural behaviors such as foraging, nest building, or suckling their young ("Going Veggie for Animals" n.d.). Similarly, the page notes that chickens, sheep, cows, and fish are reared in inhumane conditions, and those conditions are not only distressing to the animal but also can cause them to develop diseases and chronic conditions ("Going Veggie for Animals" n.d.). To prevent such pain and suffering, the individual eater has an obligation to cut back on meat consumption.

The video titled "Matt's a cat man" lays out this obligation more explicitly. The main character in the video, Matt, loves his cat, but he also loves hamburgers. The narrator draws the viewer's attention to the disconnection between Matt's affection for his cat and his meat consumption, noting that returning home to see his beloved pet after eating meat at a restaurant gave Matt "a funny feeling in his gut" ("Matt's a Cat Man" 2014). This "funny feeling" signals a discomfiting moral inconsistency, even if Matt does not recognize it initially. The narrator notes, "Cows and cats have quite a lot in common," and states that Matt had a "eureka" moment one day that illuminated his obligation to all animals, not just his cat ("Matt's a Cat Man" 2014). Afterward, Matt adopts a flexitarian diet, which helps reduce slaughter of animals and resolves Matt's moral discomfort ("Matt's a Cat Man" 2014). This video notes that humans sometimes ignore the similarities between different kinds of animals as members of the global community. Eating meat, the video argues, does not mean that a meat eater does not care about animals at all. Instead, it acknowledges that individuals may have distinct relationships with different kinds of animals, and that, as a society, we have been conditioned to view certain animals as less worthy of our care than others. The video, thus, seeks to extend the non-vegetarian's obligation of care beyond pets and outward to animals often associated only with food, even if he does not cut meat completely from his diet.

Additionally, where the Victorian Society highlighted the harm wrought on the environment and economy by the meat industry, the contemporary Society sharpens its emphasis on the obligations vegetarians and non-vegetarians have to the environment, particularly to reducing the effects of climate change. The website notes that a vegetarian diet is "one of the best things you can do to stop climate change" ("A Vegetarian Diet Is Better for the Environment" n.d.). It raises alarms about projected population gains—noting that the world population will likely increase to 9 billion people by 2050—and asserts that "we are consuming the planet's natural resources faster than the Earth can replenish them" ("A Vegetarian Diet Is Better for the Environment" n.d.). Currently, the page suggests, the world's food habits endanger the Earth and the people and animals that live on it. The website notes that a plant-based diet produces less emissions, saves water, and reduces land use ("A Vegetarian Diet Is Better for the Environment" n.d.). Therefore, to preserve a harmonious relationship with the environment, individuals have a responsibility to reduce their meat consumption.

The final video on the Society's website foregrounds these obligations. "Winnie Loves Adventure" discusses Winnie, an older woman who has travelled the world ("Winnie Loves Adventure" 2014). After researching climate change, Winnie worries that her grandchildren will not be able to enjoy the Earth in the same way. The video contrasts an image of Winnie cheerfully trekking up a mountain with an image of her granddaughter standing in a forest of stumps ("Winnie Loves Adventure" 2014). It notes that livestock cause "more pollution than the world's entire transport system" and that farming consumes numerous resources, including fresh water ("Winnie Loves Adventure" 2014). Learning this information, when cooking, Winnie switches out her sausages for beans and tomatoes because she knows that this change will use less "water and land" and "help to protect the planet from harmful greenhouse gasses." Winnie's choices mean "that everyone [can] have the chance at a great adventure" ("Winnie Loves Adventure" 2014). This video emphasizes the importance of reducing individual meat consumption for preserving the environment. Individuals not only have an obligation to protect the Earth for the Earth's sake; rather, they have an obligation to protect the Earth for the sake of other humans.

Conclusion

Though separated by over a century, the Victorian and contemporary vegetarian movements make similar arguments in favor of a vegetarian diet. Both highlight its health benefits. Both note moral imperatives about animal cruelty and concern about the effects of the meat industry on the environment and consumers. Both articulate a desire for harmony between humans, animals, and the environment. Both share a utopic vision of a world peopled by vegetarians. The contemporary and historical Societies diverge, however, on how an improved self and society can be achieved. The Victorian Society argued that individuals needed to abstain completely from meat. Conversely, the contemporary Society adopts a more permissive set of suggestions about how individuals can improve their lives and move toward a more moral and responsible world. Simply cutting back on meat—rather than removing animal products completely—can help an individual feel better, act more in line with their values, and contribute to the betterment of their world.

We can account for this difference by returning to the question of citizenship and how citizen actions can be complicated by contextual discussions of personhood and agency. As noted earlier, eating an animal can limit or contribute to an individual's recognition as a citizen of the vegetarian community. In both historical and contemporary circumstances, eating meat excludes individuals from being "vegetarians." According to Simpson and other writers for the *Vegetarian Messenger*, a carnivorous diet rendered individuals violent or unable to think clearly, leaving them with compromised agency and less able to contribute to the betterment of civic society. Thus, the *Vegetarian Messenger* argued for radical behavioral changes. This rigidity resonates with nineteenth-century ideals of citizenship, where citizenship overall was defined more narrowly, intersecting with racist and imperialist discourses. The contemporary Society focuses less on perfecting individuals. Instead, the modern Society adopts a view that different kinds of people can enact rhetorical citizenship—even if they do not earn full citizenship as "vegetarians." To benefit the whole of society, they are more willing to accept incremental, rather than radical change. Discouraging individuals from enacting citizen behaviors, even if they do not earn the full designation of "vegetarian," will ultimately do more harm than good to the cause they champion. Thus, the contemporary Society emphasizes that, as citizens of global society—vegetarian or not—individuals have obligations to the environment, animals, and other humans. They can fulfill these obligations by reducing the amount of meat in their diets, even if they do not eliminate it completely.

Although the contemporary Society adopts a more egalitarian approach, the effectiveness of this approach is far from settled. On its face, this incremental strategy encourages vegetarian behaviors, and in recent years, the number of individuals identifying as vegetarian or vegan has increased, and flexitarianism has grown in popularity.[13] Some vegetarian- and vegan-adjacent organizations, however, have argued that such measures do not go far enough—especially in the face of climate change. People for the Ethical Treatment of Animals, for example, featured an article on its website titled "The Amazon Rainforest Is Aflame, and Flexitarians Want a Cheat Day?" The author, Katherine Sullivan, lauds flexitarians for taking "a step in the right direction" but argues that such small steps are not enough, asserting, "Our planet can't settle for flexitarianism, so neither can we" (n.d., n.p.). Sullivan is not alone in noting the insufficiency of incremental changes. As Hans Dagevos writes, the increasing trend away from meat-heavy diets in high-income countries does not mitigate the rising number of meat eaters in developing nations, where meat consumption is directly correlated to the rise of the middle class (2021, 531). While the contemporary Vegetarian Society may be gradually winning individuals over to diets including less meat with their more inclusive approach, the pressing realities of climate change mean that incremental approaches may not be enough to secure the future of our society and our planet for vegetarians and non-vegetarians alike.

Notes

1 Mid-nineteenth-century periodicals such as the *Vegetarian Messenger* often elided the names of individual authors on articles. Thus, when the writer or speaker is clear, I have indicated the author. When the author is unidentified or unclear, I have not indicated an author in my in-text comments or bibliography.

2 "Veganism" was not distinguished from vegetarian until the mid-twentieth century (Cole 2010, 279). Therefore, in discussions of the nineteenth-century movement, "vegetarianism" refers to individuals who would identify as "vegetarians" and "vegans."

3 In December 1849, Simpson declares that, for "the development of all [man's] faculties to their fullest extent" the individual must "subsist on the direct productions of the vegetable kingdom, and totally abstain from the flesh and blood of the animal creation" ("The Vegetarian Principle. Lecture I: Introductory, Physical Advantages" 1849, 18). In June 1851, Simpson writes that the observance of vegetarian principles "is a stepping stone to a higher stage of existence, and removes obstructions which hinder the fulfillment of man's highest aspirations, and it is the inlet to a new and holier life" ("The Vegetarian Movement in America" 1951, 42).

4 Simpson's views of the London markets, which Ronald Morrison has termed a "noisy, bloody spectacle," was not uncommon (2015, 32). As Troy Philpotts has noted, the market kept cattle in "nightmarish" conditions: bodies lashed and bloody, horns torn off, and prone to uncontrollable "frenzy" (2010, 27). A Royal Commission recommended that the market be relocated in 1849 (Philpotts 2010, 27). In the same period as Simpson criticized "live" markets, journalists, including Charles Dickens, condemned the markets and asserted that they should be moved out of London or shut down completely (Morrison 2015, 31).

5 See "Flesh-eating and Drowsiness" 1850; "Flesh-eating and Inflammation" 1850; "Flesh-eating and Spasms" 1850; "Flesh-eating and Palpitation of the Heart" 1850; "Flesh-eating and Dyspepsia" 1850; and "Flesh-Diet and Bad Temper" 1850.

6 See "Consumption and Flesh-Diet" 1850; "Scurvy Cured by Vegetarian Diet" 1850; "Mental Vigour and Vegetable Diet" 1850; "Vegetarian Practice versus Disease" 1850; "Vegetarian Habits versus Cholera" 1850; "Asthma" 1850; "Notices to Correspondents" 1850.

7 Food-borne diseases resembling trichinella have plagued humans for centuries (Neghina et al. 2012, 504). Trichinellam was first identified in 1835 (Neghina et al. 2012, 504). Over the following decades, scientists studied the disease, linking it to pigs in 1845 and connecting it to the appearance of larvae in the body in 1850 (Neghina et al. 2012, 505).

8 Food adulteration was alarmingly common during the Victorian period. As Spencer notes, "There was little, in fact, that could be eaten and drunk which was not adulterated. Pickles were coloured green from copper, pepper had sweepings from factory floors in it, wine contained bitter almonds or spoiled cider, Gloucester cheese was coloured with red lead. By the middle of the century an excessive degree of competition between bakers, publicans and grocers led them to make greater cost cuts by even more ingenious methods of adultering the food" (1995, 265).

9 See, for example, "Health of Scottish Children" 1850; "Longevity of the Scotch" 1850; "Notices to Correspondents" 1850, 48.

10 A Mr. Gammage similarly warned of the dangers to the British from meat-eating in 1857. The "passion for indulging in flesh[,]" Gammage wrote, is both "an evil in itself" and "is promotive of other appetites equally evil" (1857, 122–3). This "evil," Gammage argued, meant that "thousands of men and women in the British Isles are, in point of intellectual status, but slightly raised above the level of the brutes which they consume" (1857, 122). Meat-eating corrupted an individual's natural intelligence to such a degree that, in Gammage's estimation, "[w]hen placed in contrast with their more advanced fellow mortals, one might suppose them to be of a different order" (1857, 122).

11 As the century progressed, Cannon Schmitt notes, these theories were more fully fleshed out by writers such as T.H. Huxley, E.B. Tylor, and Charles Darwin (2009, 34).

12 Contemporary scholarship also connects meat eating and slavery. According to Bob Torres, in an economy where animals are used as both food and agents of labor, animals are "super exploited living commodities"; lacking the ability to distinguish between work and home, animals are "always in the grasp of productive capital" (2007, 42).

13 A June 2020 Euromonitor report places the number of vegetarians and vegans at 10 percent globally and notes that 42 percent of individuals surveyed restrict but do not completely exclude animal products from their diets (Mascaraque 2020, 4).

Bibliography

"Accrington Vegetarian Association Lecture." 1855. *Vegetarian Messenger* 6 (72): 49–56.

"Alan Wishes He Loved His Waistline." 2014. YouTube video, October 1, 2020. https://www.youtube.com/watch?time_continue=1&v=cv4GhMX0bpQ&feature=emb_logo.

Asen, Robert. 2004. "A Discourse Theory of Citizenship." *Quarterly Journal of Speech* 90 (2): 189–211. doi:10.1080/0033563042000227436.

"Asthma." 1850. *The Vegetarian Messenger Supplement* 1 (12): 24.

Chávez, Karma R. 2015. "Beyond Inclusion: Rethinking Rhetoric's Historical Narrative." *Quarterly Journal of Speech* 101 (1): 162–72.

Cole, Matthew. 2010. "Veganism." In *A Cultural Encyclopedia of Vegetarianism*, edited by Margaret Puskar-Pasewicz. Santa Barbara, CA: ABC-CLIO.

Conley, Carolyn A. 2005. "Wars among Savages: Homicide and Ethnicity in the Victorian United Kingdom." *Journal of British Studies* 44 (4): 775–95.

"Consumption and Flesh-Diet." 1850. *The Vegetarian Messenger Supplement* 1 (3): 5.
Dagevos, Hans. 2021. "Finding Flexitarians: Current Studies on Meat Eaters and Meat Reducers." *Trends in Food Science & Technology* 114 (August): 530–39. doi:10.1016/j.tifs.2021.06.021.
DeLuca, Kevin Michael. 1999. "Unruly Arguments: The Body Rhetoric of Earth First!, Act Up, and Queer Nation." *Argumentation and Advocacy* 36 (1): 9–21.
"Diet in Relation to the Mental and Moral Powers" 1854. *The Vegetarian Messenger* 5 (61): 101–5.
"Experience of Vegetarians." 1850. *The Vegetarian Messenger Supplement* 1 (3): 6.
"Experimental Philosophy." 1850. *The Vegetarian Messenger* 1 (5): 58–9.
Falvo, Kathryn R. 2019. "On Apples and Anarchy." *Gastronomica* 19 (1): 33–44. doi:10.1525/gfc.2019.19.1.33.
"Ferocity and F[l]esh-eating." 1850. *The Vegetarian Messenger Supplement* 1 (5): 20.
"Flesh-Diet and Bad Temper" 1850. *The Vegetarian Messenger Supplement* 1 (3): 5.
"Flesh-eating and Drowsiness." 1850. *The Vegetarian Messenger Supplement* 1 (3): 5.
"Flesh-eating and Dyspepsia." 1850. *The Vegetarian Messenger Supplement* 1 (3): 5.
"Flesh-eating and Inflammation." 1850. *The Vegetarian Messenger Supplement* 1 (3): 5.
"Flesh-eating and Its Concomitants." 1855. *The Vegetarian Messenger* 6 (65): 21–2.
"Flesh-eating and Palpitation of the Heart." 1850. *The Vegetarian Messenger Supplement* 1 (3): 5.
"Flesh-eating and Spasms." 1850. *The Vegetarian Messenger Supplement* 1 (3): 5.
"Fruits and Farinacea the Proper Food of Man." 1850a. *The Vegetarian Messenger* 1 (4): 39–45.
"Fruits and Farinacea the Proper Food of Man." 1850b. *The Vegetarian Messenger* 1 (8): 87–91.
Gammage, Mr. "Vegetarianism and the Improvement of Society." 1857. *The Vegetarian Messenger* (8) 96: 121–6.
"Going Veggie for Animals | The Vegetarian Society." n.d. Vegetarian Society. Accessed September 27, 2020. https://vegsoc.org/info-hub/why-go-veggie/animals/.
Gregory, James. 2007. *Of Victorians and Vegetarians: The Vegetarian Movement in Nineteenth-Century Britain*. London: Tauris Academic Studies.
"Health and Nutrition | The Vegetarian Society." n.d. Vegetarian Society. Accessed September 27, 2020. https://vegsoc.org/info-hub/health-and-nutrition/.
"Health of Scottish Children." 1850. *The Vegetarian Messenger Supplement* 1 (10): 19.
"Introduction." 1849. *The Vegetarian Messenger* 1 (1): 1.
Jorgensen, Beth. 2015. "To Meat or Not To Meat?" *Poroi: An Interdisciplinary Journal of Rhetorical Analysis & Invention* 11 (1): 1–19. doi:10.13008/2151-2957.1220.
Kim, Haejoo. 2021. "Vegetarian Evolution in Nineteenth-Century Britain." *Journal of Victorian Culture* 26 (4): 519–33. doi:10.1093/jvcult/vcab040.
Kock, Christian, and Lisa Villadsen. 2014. "Introduction: Rhetorical Citizenship as a Conceptual Frame: What We Talk about When We Talk about Rhetorical Citizenship." In *Contemporary rhetorical citizenship*, edited by Christian Kock and Lisa Villadsen. 9–26. Leiden: Leiden University Press.
LeMesurier, Jennifer Lin. 2014. "Somatic Metaphors: Embodied Recognition of Rhetorical Opportunities." *Rhetoric Review* 33 (4): 362–80.

"Longevity of the Scotch." 1850. *The Vegetarian Messenger Supplement* 1 (10): 19.
Mascaraque, Maria. 2020. "Going Plant-Based: The Rise of Vegan and Vegetarian Food." *Passport: Euromonitor International.* Accessed January 31, 2022. https://go.euromonitor.com/rs/805KOK719/images/Rise_Vegan_Vegetarian_Food.pdf?mkt_tok=ODA1LUtPSy03MTkAAAGCUYJ5FYtx54SRE-oGh1AW2LrUAmzbHKLylHDzt8sBAvSelzp_sGizq0y2pi6VOIzX76SO_ZzTJP8DvUsmA9xjYoEKC1vQD7_Fx3TU0e8aehI9ibm0.
"Matt's a Cat Man." 2014. YouTube video, October 1, 2020. https://www.youtube.com/watch?time_continue=13&v=s891CczNFvo&feature=emb_logo.
"Mental Vigour and Vegetable Diet." 1850. *The Vegetarian Messenger Supplement* 1 (3): 5.
Micheletti, Michele, and Dietlind Stolle. 2012. "Sustainable Citizenship and the New Politics of Consumption." *The ANNALS of the American Academy of Political and Social Science* 644 (1): 88–120.
Morrison, Ronald. 2015. "Household Words and the Smithfield Controversy." *Victorians: A Journal of Culture and Literature*, no. 127: 31–45.
Morton, Timothy. 1998. "The Pulses of the Body: Romantic Vegetarianism and Its Cultural Contexts." *1650-1850: Ideas, Aesthetics, and Inquiries in the Early Modern Era* 4: 53–87.
NatSen Social Research. 2019. "Food and You Survey: Wave 5 Secondary Analysis: The Current Food Landscape across England, Wales and Northern Ireland." Accessed October 1, 2020. https://www.food.gov.uk/sites/default/files/media/document/food-and-you-wave-5-secondary-analysis-the-current-food-landscape-across-england-wales-and-northern-ireland-wales-final-004.pdf
Neely, Michelle C. 2013. "Embodied Politics: Antebellum Vegetarianism and the Dietary Economy of Walden." *American Literature* 85 (1): 33–60.
Neghina, Raul, Roxana Moldovan, Iosif Marincu, Crenguta L. Calma, and Adriana M. Neghina. 2012. "The Roots of Evil: The Amazing History of Trichinellosis and Trichinella Parasites." *Parasitology Research* 110 (2): 503–8.
"Notices to Correspondents." 1850. *The Vegetarian Messenger* 1 (5): 48.
Philpotts, Trey. 2010. "Mad Bulls and Dead Meat: Smithfield Market as Reality and Symbol." *Dickens Studies Annual* 41: 25–44.
"The Preying upon Animals the Trainer for War." 1855. *The Vegetarian Messenger* 6 (68): 43–6.
Puskar-Pasewicz, Margaret. 2004. "'For the Good of the Whole': Vegetarianism in 19th-Century America." PhD diss., Indiana University.
Puskar-Pasewicz, Margaret. 2010. "Preface." *Cultural Encyclopedia of Vegetarianism*, edited by Margaret Puskar-Pasewicz. vii–ix. Santa Barbara, CA: Greenwood.
Ray, Angela G. 2007. "The Rhetorical Ritual of Citizenship: Women's Voting as Public Performance, 1868–1875." *Quarterly Journal of Speech* 93 (1): 1–26.
Samek, Alyssa A. 2017. "Mobility, Citizenship, and 'American Women on the Move' in the 1977 International Women's Year Torch Relay." *Quarterly Journal of Speech* 103 (3): 207–29.
Schmitt, Cannon. 2009. *Darwin and the Memory of the Human.* Cambridge, UK: Cambridge University Press.
"Scurvy Cured by Vegetarian Diet." 1850. *The Vegetarian Messenger Supplement* 1 (3): 5.

Simons, John. 2010. "United Kingdom." In *A Cultural Encyclopedia of Vegetarianism*, edited by Margaret Puskar-Pasewicz. 229–37. Santa Barbara, CA: ABC-CLIO.
"The Slaughter Houses of Buenos Ayres" 1854. *The Vegetarian Messenger* 5 (57): 64–8.
"Social and Moral Responsibility." 1852. *The Vegetarian Messenger* 3 (31): 35–6.
Spencer, Colin. 1995. *The Heretic's Feast: A History of Vegetarianism*. Lebanon, NH: University Press of New England.
Sullivan, Katherine. n.d. "The Amazon Rainforest Is Aflame, and Flexitarians Want a Cheat Day?" Accessed January 31, 2022. https://www.peta.org/features/what-is-a-flexitarian/.
"Take Up Your Freedom." 1850. *The Vegetarian Messenger Supplement* 1 (9): 17.
Taylor, Jenny Bourne and Sally Shuttleworth. 1998. *Embodied Selves: An Anthology of Psychological Texts 1830–1890*. Oxford, UK and New York, NY: Oxford University Press.
Torres, Bob. 2007. *Making a Killing the Political Economy of Animal Rights*. Edinburgh, UK: AK Press.
"A Vegetarian Diet Is Better for the Environment | The Vegetarian Society." n.d. Vegetarian Society. Accessed September 27, 2020. https://vegsoc.org/info-hub/why-go-veggie/environment/.
"Vegetarian Habits versus Cholera." 1850. *The Vegetarian Messenger Supplement* 1 (3): 3–4.
"The Vegetarian Movement in America." 1851. *The Vegetarian Messenger Supplement* 2 (20): 42–4.
"The Vegetarian Movement. Its Claims upon Public Attention." 1851. *The Vegetarian Messenger* 2 (15): 1–4.
"Vegetarian Practice versus Disease." 1850. *The Vegetarian Messenger Supplement* 1 (3): 5.
"The Vegetarian Principle. Lecture I: Introductory, Physical Advantages." 1849. *The Vegetarian Messenger* 1 (2): 17–26.
"The Vegetarian Principle. Lecture II: Economy; Intellectual and Moral Advantages." 1850. *The Vegetarian Messenger* 1 (3): 27–36.
"The Vegetarian Society." 1849. *The Vegetarian Messenger* 1 (1): n.p.
The Vegetarian Society of the United Kingdom Limited. 2019. "Group Annual Report and Financial Statements. Year ended 31 March 2019." Accessed October 1, 2020. https://vegsoc.org/wp-content/uploads/2020/03/VegSoc-AR-to-Mar-2019-FINAL.pdf.
"Winnie Loves Adventure." 2014. YouTube video, October 1, 2020. https://www.youtube.com/watch?time_continue=57&v=vmOwtmXFeJI&feature=emb_logo.
"Why Go Veggie?" n.d. Vegetarian Society. Accessed September 27, 2020. https://vegsoc.org/info-hub/why-go-veggie/.
Young, Robert J.C. 1995. *Colonial Desire: Hybridity in Theory, Culture, and Race*. Philadelphia, PA: Routledge.

6 Vegetarianism and the Rhetorical Ecology of the Moosewood Cookbook Collective

Lora Arduser

Vegetarian, vegan, keto, paleo… the growing list of dietary labels people apply to themselves and others indicates not only a stance to food but an orientation to the world. They also have implications for our understandings of health and identities. As such "veganism and vegan identity, as well as the popular and academic discourse that constructs those categories, need to be explored, understood, and challenged" (Wright 2015, 1). To provide a response to Laura Wright's call, this chapter focuses on vegetarian discourse in a specific time and place through a methodology Judy Z. Segal (2005) frames as a kairology. A kairology, according to Segal, traces dominant cultural and rhetorical themes in a history to study such moments as rhetorical opportunities. The goal of this kairology is to better understand the historical flow of assumptions and arguments for vegetarianism in a specific rhetorical ecology (Edbauer 2005; Jensen 2015a)—a 45-year period of time in the United States. Similar to the goals Robin Jensen (2015a) describes for employing an ecological approach to the rhetoric of health and medicine, here my guiding questions concern the ways in which one moment's rhetoric constitutes specific aspects of vegetarianism and flows into or resists concurrent representations.

To trace these arguments and assumptions, I focus on genre artifacts that arise within the particular rhetorical ecology. In the study genre is defined as "a complex of elements – a constellation of substantive, stylistic and situational characteristics" (Campbell and Jamieson 1978, 17) and emphasizes the situational contexts in which genres occur (Freedman and Medway 1994). To attend to this particular genre ecology (Freedman and Smart 1997), I trace rhetorical themes in the history of vegetarianism in the United States from the 1970s to 2000s through an analysis of 14 vegetarian cookbooks written by the Moosewood Collective between 1977 and 2017. I situate the cookbooks within the social and intellectual milieu of the commune cookbook genre set to offer a window into the affective, social, and political contours of a culture's relationship with food through eating and cooking. The books, created by individuals working at the Moosewood Restaurant in Ithaca, New York, a restaurant that has been in continuous operation for 45 years, act as a field narrative that

DOI: 10.4324/9781003039013-6

portrays "general or central feelings, discussions, or representations on a subject" (Moeller and Frost 2016, 1). This particular field narrative's work as a genre repertoire, a "set of genres that are routinely enacted by the members of a community" (Orlikowski and Yates 1994, 542), can help us to better understand the shifting and stable discourses, motivations, and ideologies of veg(etari)anism in the United States over a specific period of time. In the next section, I outline how these discourses and ideologies persist or wane over time.

Kairologies and Rhetorical Ecologies

A kairology is a "robust framework for illuminating connections between history, rhetoric, and culture" (Koerber 2018, 5) that enables a rhetorician to understand how, when, and why certain arguments are persuasive (Kessler 2016). Segal (2005) uses a concept of *kairos*, the root word of kairology, to provide a rhetorical history of biomedicine—what she describes as "the 'official' medicine of the Western world" (23). Following Segal, other scholars in rhetoric and composition and in the rhetoric of health and medicine have used the framework to theorize a rhetoric of wearability (Kessler 2016), the rhetorical activity preceding the emergence of the word "hormone" as a way of understanding women's health (Koerber 2018), and cancer rhetoric's development in the United States (Agnew 2018). And while these projects operate in the realm of the rhetoric of health, medicine, and science, a kairological approach is appropriate for other rhetorical/historical projects because, as Segal importantly points out, what a rhetorician is interested in is "not only a *chronology* of events but also a *kairology* of them: a study of historical moments as rhetorical opportunities" (23 [emphasis in original]).

To illuminate the connections and shifts in arguments for vegetarian discourse, I emplace this kairology in a particular rhetorical ecology (Edbauer 2005) using a percolation model (Jensen 2015a) to trace the communication of ideas about vegetarianism in a particular setting and time period. Edbauer productively complicates models of public communication (i.e., the sender-receiver-text triangle and Bitzer's model of the rhetorical situation as well as critiques and modifications of this model) with the concept of rhetorical ecologies. Building from these critiques and modifications, she "look[s] towards a framework of *affective ecologies* that recontextualizes rhetorics in their temporal, historical, and lived fluxes" (Edbauer 2005, 9 [emphasis in original]). In doing so, like Barbara Biesecker (1989), Louise Phelps (1988), and Michael Warner, (2002), Edbauer adds the dimensions of history and movement into rhetorical situations.

Jensen takes up the concept of rhetorical ecologies to discuss the "inherently and inescapably intra-and-inter-disciplinary rhetoric of health scholarship" (2015a, 523). Her interest is in heeding the processes "through which health-related ideas, assumptions, and arguments have been communicated by and in relationship with the technical sphere, public or mainstream audiences, counter-publics, and vernacular or lay constituencies" (2015a, 523), and she offers two models to do so within rhetorical ecologies: the flow or circulation model and the percolation model. The flow/circulation model traces the communication of ideas, assumptions, and arguments chronologically, such as Jensen's own work (2015b) on the medicalization of infertility in the 1930s and 1940s. The percolation model, based on the philosophy of Michel Serres (1991, 1995), makes connections between rhetoric in different, distinct time periods and highlight that the ideas, assumptions, and arguments of particular historically distinct moments can and do re-emerge and repeat themselves in these different time periods. In the following analysis of the Moosewood collection, I detail the assumptions and arguments with which the Moosewood texts engage.

The Rhetorical Ecology of the Moosewood Cookbooks

As Janet Floyd and Laurel Forster note in *The Recipe in its Cultural Contexts*, cookbooks and the recipes between cookbook covers "evoke the elaborate scene of home, and the contentious arena of domestic politics and family values. In their different appearances, they are also persistently drawn into cultural debates around health and purity, about lifestyle and individualism, and into definitions of the national past, present, and future" (2003, 1). Community cookbooks specifically capture and preserve the tastes and voices of their various authors, churches, auxiliaries, athletic teams, neighborhoods, and civic organizations. Scholarly research on community cookbooks have focused on those written by groups of women, with the analysis frequently focusing on examining how authors "subvert[ing] the subversive roles into which they were born incorporating food as a method of resistant" (Ferguson 2012, 697). Using the Lutheran Church Women cookbook as a case study of the traditions of female collectivity through cookbook publishing, Ferguson notes the two perspectives generally brought to scholarship on such cookbooks: as "natural outgrowths of politico-societal organizations" (2012, 697) or as opportunities for the authors to subvert traditional gender roles, using cooking as a form of resistance. In other words, the focus of "collective" has been on gender roles and how these cookbooks might function as oppressive or emancipatory for the women compiling and publishing them. But as Ferguson (2012) also points out, community cookbooks offer an opportunity to examine how the books function as a process of community building.

Commune cookbooks were also interested in the process of community building. Unlike the community cookbooks discussed by Ferguson, however, commune cookbooks like the one published by The Farm (1975) were more interested in the agrarian lifestyle historically embraced by communes in the United States in the 1960s and 1970s (Hartman 2003). In her study of commune cooking, Stephanie Hartman specially places the first Moosewood cookbook alongside these "haired predecessors, with their instructions on cooking with psychedelics, making pemmican sausage, preserving one hundred pounds of "web-stomped" sauerkraut, concocting medicinal teas, and developing a rapport with vegetables" (2003, 32).

The Moosewood cookbook collection is situated within the genre of such commune cookbooks, even if less "haired." Although the Moosewood Collective has been fluid in the 46 years it has been operating, according to the restaurant's website (2019), it is self-described as a group of 19 people who come from various occupations and includes musicians, singers, activists, teachers, writers, gardeners, editors, poets, artists, quilters, and health advocates. From the 1970s on, the Moosewood cookbooks produced by this collective retains a nonradical arm of vegetarianism and argue for vegetarian as healthy eating, cooking, social pleasure of eating, appreciation of local and global, and farm to table versus "big agriculture."

Moosewood and Vegetarianism's Counterculture Roots

With three recessions, double-digit inflation, and double-digit unemployment, the 1970s were a time of political upheaval, largely associated with the Vietnam War. Facing an uncertain future, some Americans searched for meaning through alternative lifestyles such as communal living. Megan Elias writes that this communal way of life "was a notable phenomenon in the Vietnam War era, largely practiced by young middle-class European Americans, usually college graduates. Scholars of the movement saw it as a rejection of middle-class consumerist culture" (147). Estimates of the number of communes in the United States in the early 1970s range from 2,000 to 3,000 (Elias 2017). As a part of their lifestyle, members of communes embraced vegetarianism and cookbooks published by these communes expressed the group's rejection of mainstream, middle-class values through critiquing and politicizing industrial food practices and the practice of eating meat. Cookbooks filled with recipes like the ones Lucy Horton collected from 43 communes for her cookbook *Country Commune Cooking* (1972) increased in the 1970s: Between 1940 and 1970, approximately 15 natural foods cooking books were published in America; between 1970 and 1980, the number exceeded 300 (Elias 2017).

An important voice in these publishing efforts was The Farm, a religious vegetarian community in Tennessee. Started by Stephen Gaskin, an English professor from San Francisco State University, in 1970, the members of the commune grew their own food and even had their own phone company, healthcare system, and water system. The Farm started with 250 people but 1980 consisted of approximately 1,100 people living on 1,750 acres of farm land that included 250 acres of soybeans (Iacobbo and

Iacobbo 2004). The residents of the Farm advocated for vegetarianism as a way to help world hunger: "We are vegetarians because one-third of the war is starving and at least half goes to bed hungry every night. If everyone was vegetarian there would be enough food to go around, and no one would be hungry" (The Farm 1975, 5).

Along with world hunger, vegetarians in the 1970s voiced concerns about the ethical treatment of animals. Using TV to make these arguments, Gypsy Boots, author of *Bare Feet and Good Things to Eat* and owner of the restaurant Back to Nature Health Hut, made regular appearances on the Steve Allen Show to promote vegetarianism as both a healthy lifestyle and lifestyle that helped alleviate the suffering of animals bond for the slaughterhouses. Other arguments for vegetarianism in the 1970s resonated with the one made in 1962 by biologist Rachel Carson in *Silent Spring*. Arguably, one of the most influential books of the twentieth century, the text helped spark the modern ecology movement and encouraged readers to distrust mainstream American agriculture (Elias 2017, 148). Within cookbooks specifically, Adelle Davis took up the same cause, prefacing the revised edition of her 1947 *Let's Cook It Right* (1962) with the ominous note that the food industry had been quietly adding chemicals to our food for the past 15 years.

In 1977, Moosewood joined other vegetarian publishing efforts with their first cookbook. Unlike more politically driven efforts, Iacobbo and Iacobbo (2004) characterize the Moosewood cookbooks as being "more interested in cuisine than economics or ecology" (195). Starting with this initial cookbook and continuing over the decades, the Moosewood Collective—like Edward Espe Brown, the American Zen teacher and author of *The Tassajara Bread Book,* and Alice Water, owner of the famous Chez Panisse in Berkeley and the author of *Chez Panisse Menu Cook Book*—believed "virtue and pleasure could exist on the same plate" (Elias 2017, 172). But between the covers of the Moosewood books, little ink is devoted to traditional vegetarian principles or "virtues" of addressing world hunger or treating animals ethically. The hand-lettered, hand-illustrated 1977 book is organized into typical cookbook sections: Introduction, Pantry Notes, Some Useful Tools, Table of Conversions, Soups, Salads, Sauces and Dips, A Few Baked Things and Sandwiches, Entrees, and Desserts. Mollie Katz writes in the book's preamble to the 40th anniversary edition:

> The Moosewood Cookbook grew, in part, out of a looseleaf binder filled with random notes intended to help track what we were cooking in the tiny kitchen of our modest 1970s restaurant. With neither chef not set menu, we were a group of friends (none of us culinary professionals taking turns in the kitchen, cooking heartfelt versions of the food we loved, beginning with family favorites.
>
> (Katzen 2014, ix)

Rather than stressing the treatment of animals, the book focuses more on treating people well:

> We fret and sometimes doubt ourselves, hoping always to find safe harbor. We've tried to live up to the expectations people have invested in our image. Sometimes we fall short and sometimes we resist being the least bit famous. Most of us experience the Collective as a kind of second family. We've shared a significant portion of our lives with one another and find Moosewood a loving and supportive place that can drive us crazy too. Sometimes it's easy, sometimes it's hard … having so much good food around sure doesn't hurt.
>
> (The Moosewood Collective 2001, 494)

Although the Moosewood cookbooks do not fully echo the political concerns of other communal cookbooks of this period, the attention to family displayed in these openings are reminiscent of the communal lifestyle embraced by communes and expressed in Commune cookbooks like The Farm's:

> We have been on the land for seven years as a community and have migrated to Tennessee from San Francisco. Original group came together through open meetings held by Stephen, who is the Farm's founder and minister. We are vegetarians because 1/3 of the world is starving and at least half goes to bed hungry every night. If everyone was vegetarian, there would be enough food to go around, and no one would be hungry. (5)

For Moosewood, family is more localized than the global family The Farm refers to. For Moosewood at this point in time, the extension of family is more localized—the restaurant and Ithaca community. Through these texts both The Farm and Moosewood question the primacy put of the nuclear family, what David Brooks referred to as a "cultural ideal" in his 2020 editorial for *The Atlantic*, that began to grow in the 1970s and advocate for ideas about what might be considered an extended family.

Vegetarianism and Links to the Healthism Ideology

One of the most persistent rhetorical links to vegetarian lifestyles in the United States has been the ideology of healthism. Robert Crawford, a political economist who is attributed with coining the term, defined healthism as the "preoccupation with personal health as primary—often the primary—focus for the definition and achievement of well-being." (1980, 368). The interest in these principles of healthism was strong in the 1970s when the American healthcare system was a critical concern. While the introduction of imaging technologies such as magnetic resonance imaging

(MRI) had the potential to extend and improve the quality of people's lives, these were also part of an expensive and often ineffective medical system (Crawford 1977). As such, access to care was also a top political concern in the 1970s. These concerns were evident in the competing national health insurance proposals that came from both senators including Senator Ted Kennedy and Presidents Richard Nixon and Gerald Ford (Kaiser Family Foundation 2011).

During the 1980s, vegetarians remained outspoken about the ethical treatment of animals, as evidenced by the founding of People for the Ethical Treatment of Animals (PETA) and the Farm Sanctuary and the first ever sanctuary for farmed animals. The continuing concern for politically driven concerns with the destruction of the environment and animal liberation also continued strong in the 1980s. Originally published in 1987, for example, *Diet for a New America* encouraged readers to take a hard look at the food they were buying and eating. Often seen as the bible of vegetarianism, the book focuses on the line connecting America's factory farm system with disease, animal cruelty, and the ecological crisis.

In this decade, vegetarianism also continued to be closely linked to healthism, which acknowledges that even though health problems may originate outside the individual from factors such as a poor diet, the solutions to the problems can be reached through individual choice (Crawford 1980). In the 1980s, this choice focused on cardiovascular health and the relationship between eating red meat high saturated fatty acids and cardiovascular disease and some cancers. Another signal of the increased attention to personal health through dietary choices is that the first USDA dietary guidelines came out in 1980. In February of that year, the USDA and HHS collaboratively issued Nutrition and Your Health: Dietary Guidelines for Americans. The guidelines described seven principles for a healthy diet and provided guidance on limiting dietary components such as sugar, fat, saturated fat, cholesterol, and sodium (Dietary Guidelines for Americans, n.d.).

The sole cookbook from the Moosewood Collective this decade was *New Recipes from Moosewood Restaurant.* This particular book may be the best illustration that the Collective sees vegetarianism as "one of life's sweet pleasures" (xv) as well as a healthy, ethical stance toward life. This cookbook came out the same year *Diet for a Small Planet* was published and 10 years after the Collective's first cookbook. The book, featuring an ideal scene of a family picnicking under a tree in a pasture with several cows in the background, is mostly black and white like the first one, but not hand-lettered. It also features several full-page color photos of people cooking and relaxing, as well as a black and white photo of the members of the Moosewood Collective. The cookbook itself is not strictly vegetarian, featuring a section on fish, but what's most interesting is the cookbook's focus on teaching people to cook. In the introduction the writers state: "Eating is one of life's sweet pleasures – so is cooking and serving good food to others" (xv).

They continue:

> More than a decade ago, we were just discovering how to be vegetarians. Soy beans and brown rice were a challenge. Yogurt was a little exotic, remember? Well, since then we found the world full of foods and flavors taste and savor. And you know, this vegetarian food is not just love by vegetarians. Many of us are not strict vegetarians and neither are most of our customers, who come to the suite for the good food.
>
> We're excited to be part of the developing American vegetarian cuisine. We think that the real inspiration for a cooking is an openness to ethnic and regional foods – and readiness to mix and match in the American melting-pot tradition, taking the best and most interesting and throwing it together in it in new ways to create something unique and wonderful. (xv)

This cookbook, therefore, embodies what Crawford (1980) describes as healthism's "broader philosophic orientation, holistic health is portrayed as 'a way of being,' an interrelation or balance of body, mind, and spirit, a concern with 'high-level wellness,' 'super health,' or the 'joy of life'" (p. 366). The enjoyment of food expressed by Moosewood also includes the introduction of new dishes and exotic ingredients; therefore, ethnic dishes, dishes that had been a staple of vegetarian cooking since the 1970s, remain a focus of both vegetarian eating and Moosewood's recipes. As noted in the 1975 Farm cookbook, ethnic cuisines "adapt easily to vegetarian cooking and add a spicy variety to our meals" (6). The Moosewood Collective regularly incorporated recipes like hot and sour soup (from 1977), Japanese Fish Stew (from 1987), but its own ethnic-food focused cookbook, *Sundays at Moosewood*, did not come out until 1990. The introduction in the book acknowledges that the interest in international cooking as well as eating has remained strong over the decades, but by the 1990s, hard-to find ingredients were also becoming more accessible. Still, unlike earlier vegetarian cookbooks, Moosewood continues to link its choices to joy and pleasure and well as health and ethical values:

> Today, cuisines are changing faster and faster, and at Moosewood we're happy to be a part of that process. At the same time, the preparing and serving of foods is one of the most interesting and carefully preserved identifiers of a culture. Traditional food is a balm for the uncertainties and confusion of a new life; the most usual common bond among immigrants it's the sharing of food. At Moosewood, all of us have discovered time and again the in integral connection that food has culture, heritage, the returning of history through family stories, and the celebration of rituals and seasons of the year. (11)

The authors continue that it is "healthy and exciting to foster an appetite" (11) for other cuisines, but in these statements, the Collective clearly articulates health to a connection to the world and to family, rather than the absence of disease. As such, ideas about family continue to percolate in the evolving rhetorical ecology of vegetarianism.

Along with a sustained interest in global cuisines, the 1990s witnessed increasing advocacy for vegetarianism based on the rise of more scientific evidence that supported the healthfulness of not just vegetarian diets but vegan diets. In this decade, The American Dietetic Association officially endorses vegetarianism and books by prominent doctors promote low-fat vegan or mostly vegan diets (e.g., *The McDougall Program* and *Dr. Dean Ornish's Program for Reversing Heart Disease*). The U.S. government also replaces the meat- and dairy-industry-sponsored Four Food Groups with a Food Pyramid in 1992, showing that most of a person's diet should be based on grains, vegetables, beans, and fruits.

As cookbook writers and publishers, the Moosewood Collective published a total of six cookbooks in this decade:

1. *Sundays at Moosewood Restaurant*
2. *Moosewood Restaurant Cooks at Home*
3. *Moosewood Cooks for a Crowd*
4. *Moosewood Restaurant Low-Fat Favorites*
5. *Moosewood Restaurant Book of Desserts*
6. *Moosewood Restaurant Daily Special*

Much as medical advice about eating healthy had changed over the years, as the preface of the 1996 *Moosewood Restaurant Cooks for a Crowd* notes, the audience for the restaurant and the cookbooks changed. The writers attribute this transition to an increasing interest in meatless meals regardless of whether or not a person was a vegetarian. They saw this interest being the result of a variety of factors: the awareness of the connection between food and health, a greater appreciation of global vegetarian dishes, and a concern for sustainability. In other words, "it's no longer weird to eat this way" (1996, 13). A simple comparison between the description of where to purchase black mustard seed from the 1987 *New Recipes from Moosewood Restaurant* and the 1999 *Moosewood Restaurant Daily Special* can illustrate this:

> Black mustard seeds are available in Asian markets and will keep indefinitely in the refrigerator.
>
> (1987, 283)

> Black mustard seeds are available in spice shops, Indian food stores, and many supermarkets.
>
> (1999, 379)

The locations a shopper can find black mustard seeds shift from out-of-the-way stores that may be difficult to find to the grocery store people visit regularly. The move from being weird to a widely accepted way of eating for health reasons, is evidenced in the content of *Moosewood Restaurant Cooks at Home*, *Moosewood Cooks for a Crowd*, *Moosewood Restaurant Low-Fat Favorites*, and *Moosewood Restaurant Daily Special* in other ways as well. Whereas early books would espouse the health benefits of a vegetarian diet, the recipes in all these books as well as the *Moosewood Restaurant Book of Desserts* rely on scientific data through nutritional analyses for each recipe to alert the readers to the specific nutrients or lack of unhealthful ingredients. However, the books from this time frame also resist the trends of mainstream vegetarianism of this decade as largely if not completely for health reasons. This resistance, what Elizabeth Driver (2009), would have referred to as the authors' attitude toward the food in their recipes, can often be found in non-recipe information such as a preface, the table of contents, or a dedication are places cookbook authors' express their attitudes toward the food they use in recipes (Driver 2009).

In Moosewood's cookbooks, these interests include educating people how to cook (not providing instructions to follow to create a specific dish) and, as always, making eating a pleasurable experience. These attitudes are clearest in the books' introductions and appendices. For example, the *New Recipes from Moosewood Restaurant* and *Sunday Suppers* both include an appendix entitled "What We Mean When We Say One Medium Onion." The appendix is a simple table that equals a single recipe ingredient, like a medium onion or medium tomato, with specific measurements, both in volume and weight. Other appendices, such as "If You Need to Know More," function as a dictionary of sorts, introducing readers to ingredients they may be unfamiliar with: "Black mustard seeds: Widely used in Indian cuisine, black mustard seeds have a nutty aroma and mild flavor when heated. To use, either dry-roast and grind them, or heat them whole in oil until they pop. Heat develops the flavor and inhibits the enzyme responsible for pungency" (1999, 379).

Undoubtedly, early vegetarian cookbooks from the 1970s were also interested in teaching readers to cook because stores did not carry tofu or many of the other items that have become staples in many vegetarian diets. But the ethos of this instruction was more of one of necessity than pleasure. The Moosewood cookbooks, on the other hand, linked cooking and eating directly to joy. Moosewood's ideology around the pleasure of vegetarianism is found in the introduction of *Moosewood Restaurant Cooks at Home*: "The food at Moosewood has always derived from an ethnic home-style tradition—"peasant food," created by hardworking people who cooked for themselves using minimal equipment, fresh foods, and simple cooking methods" (1994, 12). However, simplicity for Moosewood is not a contrast to pleasurable eating, and as they note, mainstream culture had begun to recognize that "The goals of a nutritious vegetarian diet and gourmet dining have drawn much closer" (1994, 13). The Moosewood appendices combine these interests with the trend of eating vegetarian for person health: These appendices list vegan and nondairy recipes, but another list has sample menus combining recipes in the books for an eating "experience," indicating a desire for a pleasurable experience as well as a healthy one. Final evidence of these seemingly incompatible values lies in the content of the books published in this decade. Along with cookbooks with vegan recipes and recipes with nutritional analyses, the Collective produced books devoted entirely to pleasurable eating as well as healthy eating, such as a dessert book, a book on ethnic cuisine, a book with recipes for special occasions and one with recipes for large groups.

Turn-of-the-Century Themes

In the ten years following the turn of the century, Moosewood published an additional five cookbooks as well as one book in 2013 and one in 2017.

1. *Moosewood New Classics*
2. *Moosewood Restaurant Celebrates*
3. *Moosewood Restaurant: Simple Suppers*
4. *The Moosewood Restaurant Kitchen Garden*
5. *Moosewood Restaurant Cooking for Health*
6. *Moosewood Restaurant Favorites* (2013)
7. *The Moosewood Restaurant Table* (2017)

Red meat and dairy were still social villains for good health in these decades, as illustrated by President Bill Clinton's adoption a vegan diet in 2010 (no meat, eggs, dairy, or other animal products) after his second heart surgery. In August 2011, in an interview he told CNN's Dr. Sanjay Gupta, "my blood tests are good, and my vital signs are good, and I feel good, and I also have, believe it or not, more energy" (Brown 2011). That same year the Food Pyramid was replaced by My Plate the same year, with Michelle Obama unveiling the image representing the five essential food groups. "My Plate" renamed the "Meat & Beans" category "Protein" and changed the "Milk" to "Dairy." The other three categories (Grains, Vegetables, and Fruits) remained the same. In the early 2000s, however, the newest health culprit was carbohydrates (Iacobbo and Iacobbo 2004). Moosewood captured this societal dietary trend in the 1999 book *Moosewood Daily Special* and the 2001 *Moosewood Restaurant New Classics* with not only a list of low-fat recipes and vegan recipes in each but a list of low carbohydrate recipes. Other vegetarian trends that the Moosewood Collective echoes in this decade include the growing interest in organic food and local food, and interest that resonates with the Slow Food Movement in the United States. This movement, originally an Italian organization, entered the consciousness of Americans and in 2000 and 2003 when Slow Food chapters were established in the United States (Schneider 2008). The philosophy behind the movement was one that promoted local food and traditional cooking. The 2001 *Moosewood Restaurant New Classics: 350 Recipes for Homestyle Favorites and Everyday Feasts* captures these interests in its introduction and appendices. The introduction to this book, like many of the Moosewood books sings the praises of simple, good food:

> Moosewood food has been called eclectic and cross-cultural because we serve dishes from so many different ethnic or regional cuisines. Moosewood often serves food prepared in a non-traditional way, but we're really good at preserving the intent, borrowing and adapting the central idea while maintaining the integrity of the original. We love to cook and are constantly improvising and experimenting. We want our vegetable sun-ripened, the cheeses savory and robust, the herbs freshly picked, the spices newly roasted and ground, and the grains whole and hearty. In a nutshell, we want flavors that sing. (14)

The appendices in the same book focus on organic growing, defining federal organic standards in 2001, and discussing GMOs.

Interestingly, the last two cookbooks produced by Moosewood to date, are similar to the others in terms of a cooking and eating philosophy, but how the books display these values is different than earlier decades. The 2013 *Moosewood Restaurant Favorites* has a much more commercial and polished production with a professional photo on the cover rather than an illustration, higher quality paper, and color photos throughout: A look that is repeated in 2017's *The Moosewood Restaurant Table*. The 2013 book also has a preface entitled "Organic Food, the Dirty Dozen, and Buying Local," but the more pleasurable side of eating is represented in the book's photos more than the text. These full-page color photos of the restaurant's food are more reminiscent of cooking magazine such as *Bon Appetite* than the earlier Moosewood's hand-drawn, black-and-white illustrations, a trend that continues in the 2017 book, which includes similar color photos, but no introductory or back matter material that so clearly defines Moosewood's ethos as instructors on eating, enjoying, and cooking in a vegetarian way. At the same time, the introduction of this volume may be the clearest line to its commune cookbook predecessors: "Many of us believe that fresh, nutritious food is a human right, and we work to support child with nutrition and family farms, increase the number of local growers, see agricultural workers make a living wage, turn the food deserts into oases with community gardens and regionally sourced food, and make healthy food affordable and accessible to all people" (1).

This quotation clearly illustrates that while embracing both the healthism individualistic appeals of personal health, the constant for the Moosewood Collective lives in the term "collective." In the 1970s Moosewood envisions it as local community, but in this decade, the rhetoric Moosewood turns to the same concerns espoused by The Farm in its cookbook from the 1970s. Through the 1970s and into the twenty-first century, the ideology of healthism has remained a stalwart of vegetarian lifestyles and "has become a powerful cultural discourse pressuring individuals to (re)frame their lifestyles around the pursuit of better health" (Silchenko and Askegaard 2020, 136). As the Moosewood cookbooks illustrate, what makes vegetarian rhetoric distinct from an ideology of healthism as expressed by Crawford (1980, 2006) is the focus on family health as well as individual.

Conclusion

The themes attached to vegetarianism in the United States—the ethical treatment of animals, the need for access to clean food, good individual health, and the importance of connection to family and globe—are all apparent in the Moosewood cookbook collection, but the rhetorical ecology Moosewood operates within also privileges abstinence and healthism as values. The broader discourse of vegetarianism in the United States has been linked with ethical questions about the environment and health in ways that privilege "doing without" by not eating meat or eliminating other items from a person's diet whether that item be gluten, carbohydrates, or diary. In many ways, these eliminations are linked to Puritanical ideas about being moral and "clean." Margaret McCartney, a medical professional, says in a short editorial in *BMJ*: "The command to eat cleanly implies that everyone else is filthy, being careless with their bodies and lives. It comes with promises of energy boosts, glowing skin, spirituality, purity, and possibly immortality. But this nonsense is all based on a loose interpretation of facts and a desire to make the pursuit of wellbeing an obsessive, full time occupation" (McCartney 2016). Ironically this quotation highlights the reverse of identities attached to vegetarianism in the 1970s, the "haired predecessors" of the Moosewood cookbooks Stephanie Hartman discusses (2003, 32) and the stereotypes of "dirty hippies" who were living communally and eating vegetarian.

Such rhetorical framings are "organizing principles that are social shared and persist over time, that work symbolically to meaningful structure the social would" (Reese 2001, 11). And although I offer chronological outline of the different themes or frames presented in the Moosewood's cookbooks, these show opportunities for the various themes within the larger culture to percolate up through the personality and regionality of the cookbook collection. This is important because as Rod Preece notes in his book *Sins of the Flesh*: "a society may best be understood less by how it answers questions and more by what questions it perennially fails to ask itself" (2008, 19). In looking at these frames through a kairology, we have a window into how stable these arguments have remained over time, even when the focus or priority shifts. However, examining the Moosewood collective, which also pushes back against stereotypes of vegetarianism in a subtle but effective way—consistently declaring over the decades that eating vegetarian can taste good as well as be good for you. This view opens opportunities for other scholars in rhetoric and food studies to have the opportunity to ask what questions we fail to ask about food, health, and identity.

Bibliography

Agnew, Lois. 2018. "Ecologies of Cancer Rhetoric: The Shifting Terrain of U.S. Cancer Wars, 1920–1980." *College English*, 80 (3): 271–96.

Biesecker, Barbara A. 1989. "Rethinking the Rhetorical Situation from Within the Thematic of Différance." *Philosophy and Rhetoric*, 22 (2): 110–30.

Bitzer, Lloyd F. 1968 "The Rhetorical Situation." *Philosophy and Rhetoric*, 1: 1–14.

Bower, Anne L. 1997. "*Our Sister's Recipes*: Exploring "Community" in a Community Cookbook." *Journal of Popular Culture*, 31 (3): 137–51.

______. ed. 1997. *Recipes for Reading: Community Cookbooks, Stories, Histories.* Amherst: University of Massachusetts Press.

Brooks, David. 2020. "The Nuclear Family Was a Mistake." *The Atlantic*. Accessed January 18, 2022. https://www.theatlantic.com/magazine/archive/2020/03/the-nuclear-family-was-a-mistake/605536/

Brown, Eryn. (2011). "Bill Clinton Talks about Being a Vegan." *Los Angles Times.* Accessed June 20, 2019. https://www.latimes.com/health/la-xpm-2011-aug-18-la-heb-bill-clinton-vegan-20110818-story.html

Campbell, Karlyn Kohrs and Kathleen Hall Jamieson. 1978. "Form and Genre in Rhetorical Criticism: An Introduction. In *Form and Genre: Shaping Rhetorical Action*, edited by Karlvn Kohrs Campbell and Kathleen Hall Jamieson, 9–32. Fall Church, VA: The Speech Communication Association.

Crawford Robert. 1977. "You are Dangerous to Your Health: The Ideology and Politics of Victim Blaming." *International Journal of Health Services*. 7 (4):663–80.

______. 1980. "Healthism and the Medicalization of Everyday Life." *International Journal of Health Services*.10 (3):365–88.

______. 2006. "Health as a Meaningful Social Practice." *Health*. 10 (4):401–20.

Davis, Adelle. 1962. *Let's Cook It Right*. New York, NY: Harcourt, Brace.

Dietary Guidelines for Americans. n.d. "History." *USDA/HHS*. Accessed May 23, 2020. https://www.dietaryguidelines.gov/about-dietary-guidelines/history-dietary-guidelines

Driver, Elizabeth. 2009. "Cookbooks as Primary Sources for Writing History: A Bibliographer's View." *Food, Culture & Society*, 12 (3): 257–74.

Edbauer, Jenny. 2005. "Unframing Models of Public Distribution: From Rhetorical Situation to Rhetorical Ecologies." *Rhetoric Society Quarterly*, 35 (4): 5–24.

Elias, Megan J. 2017. *Food on the Page: Cookbooks and American Culture.* Philadelphia, PA: University of Pennsylvania Press.

The Farm. 1975. *Vegetarian Cookbook*. Summertown, TN: The Book Publishing Company.

Ferguson, Kennan. 2012. "Intensifying Taste, Intensifying Identity: Collectivity through Community Cookbooks." *Signs*, 37 (3): 695–717.

Floyd, Janet and Laurel Forster (eds). 2003. *The Recipe in its Cultural Contexts*. New York, NY: Routledge.

Freedman, Aviva and Peter Medway, P., eds. 1994. *Genre in the New Rhetoric.* London: Routledge.

Freedman, Aviva and Graham Smart. 1997. "Navigating the Current of Economic Policy: Written Genres and the Distribution of Cognitive Work at a Financial Institution. *Mind, Culture, and Activity*, 4 (4): 238–55.

Hartman, Stephanie. 2003. "The Political Palate: Reading Commune Cookbooks." *Gastronomica*, 3 (2): 29–40.

Iacobbo, Karen and Michael Iacobbo. 2004. *Vegetarian America: A History.* Westport, CT: Praeger.

Jensen, Robin. E. 2015a. "An Ecological Turn in Rhetoric of Health Scholarship: Attending to the Historical Flow and Percolation of Ideas, Assumptions, and Arguments." *Communication Quarterly*, 63 (5): 522–26.

______. 2015b. "Improving upon Nature: The Rhetorical Ecology of Chemical Language, Reproductive Endocrinology, and the Medicalization of Infertility." *Quarterly Journal of Speech*, 101 (2): 329–53.

Kaiser Family Foundation. 2011. "*Timeline: History of Health Reform in the U.S.*" Kaiser Family Foundation. Accessed January 10, 2022. https://www.kff.org/wp-content/uploads/2011/03/5-02-13-history-of-health-reform.pdf

Koerber, Amy. 2018. *From Hysteria to Hormones: A Rhetorical History.* University Park, PA: Penn State University Press.

Kessler, Molly M. 2016. "Wearing an Ostomy Pouch and Becoming an Ostomate: A Kairological Approach to Wearability." *Rhetoric Society Quarterly*, 46 (3): 236–50.

Moeller, Marie E. and Erin A. Frost. 2016. "Food Fights: Cookbook Rhetorics, Monolithic Constructions of Womanhood, and Field Narratives in Technical Communication." *Technical Communication Quarterly*, 25 (1): 1–11.

McCartney, Margaret. 2016. "Clean Eating and the Cult of Healthism. *BMJ.* Accessed December, 20, 2021. https://www.bmj.com/content/354/bmj.i4095

Moosewood. n.d. "About." *Moosewood Restaurant.* Accessed May 22, 2019. https://moosewoodcooks.com/about/

Orlikowski, Wanda. J. and JoAnne Yates. 1994. "Genre Repertoire: The Structuring of Communicative Practices in Organizations." *Administrative Science Quarterly*, 39 (4): 541–74.

Phelps, Louise. W. 1988. *Composition as a Human Science.* New York, NY: Oxford University Press.

Preece, Rod. 2008. *Sins of the Flesh: A History of Ethical Vegetarian Thought.* Vancouver, BC: UBC Press.

Reese, Stephen D. 2001. "Prologue-Framing Public Life: A Bridging Model for Media Research." In *Framing Public Life: Perspectives on Media and Our Understanding on the Social World*, edited by Stephen D. Reese, Oscar Gandy, Jr., and August E. Grant, 7–31. Mahwah, NJ: Lawrence Erlbaum Associates.

Schneider, Stephen. 2008. "Good, Clean, Fair: The Rhetoric of the Slow Food Movement." *College English*, 70 (4): 384–402.

Segal, Judy Z. 2005. *Health and the Rhetoric of Medicine.* Carbondale, IL: Southern Illinois University Press.

Serres, Michel. 1991. *Rome: The Book of Foundations.* (F. McCarren, Trans.). Stanford, CA: Stanford University Press.

______. 1995. *Conversations on Science, Culture, and Time.* (R. Lapidus, Trans.). Ann Arbor, MI: University of Michigan Press.

Silchenko, Ksenia and Søren Askegaard. 2020. "Mapping Moralities of Food and Health in Marketing Research Literature." *Journal of Marketing Management*, 36(9–10): 794–829.

Vegetarian Society. n.d. "World History of Vegetarianism." *Vegetarian Society.* Accessed December 12, 2019. https://www.vegsoc.org/history.

Warner, Michael. 2002. "Publics and Counterpublics." *Public Culture*, 14 (1): 49–71.

Wright, Laura. 2015. *The Vegan Studies Project: Food, Animals, and Gender in the Age of Terror*. Athens, GA: University of Georgia Press.

Zaraska, Marta. 2016. "This Is Why Vegetarianism Didn't Catch on Until Recently." *Time*. Accessed February 23, 2016. https://time.com/4220270/vegetarianism-history-meathooked/.

Moosewood Bibliography

Hirsch, David. 2005. *The Moosewood Restaurant Kitchen Garden*. Berkeley, CA: Ten Speed Press.

Katzen. Mollie. 1977. *The Moosewood Cookbook*. Berkeley, CA: Ten Speed Press.

______. 2014. *The Moosewood Cookbook: 40th Anniversary Edition*. Berkeley, CA: Ten Speed Press.

The Moosewood Collective. 1987. *New Recipes from Moosewood Restaurant*. Berkeley, CA: Ten Speed Press.

______. 1990. *Sundays at Moosewood Restaurant: Ethnic and Regional Recipes from the Cooks at the Legendary Restaurant*. New York, NY: Simon & Schuster.

______. 1994. *Moosewood Restaurant Cooks at Home*. New York, NY: Simon & Schuster.

______. 1996. *Moosewood Restaurant Low-Fat Favorites*. New York, NY: Clarkson Potter/Publishers.

______. 1996. *Moosewood Cooks for a Crowd*. New York, NY: Gramercy Books.

______. 1997. *Moosewood Restaurant Book of Desserts*. New York, NY: Clarkson Potter/Publishers.

______. 1999. *Moosewood Restaurant Daily Special*. New York, NY: Clarkson Potter/Publishers.

______. 2001. *Moosewood New Classics*. New York, NY: Clarkson Potter/Publishers.

______. 2003. *Moosewood Restaurant Celebrates: Festive Meals for Holidays and Special Occasions*. New York, NY: Clarkson Potter/Publishers.

______. 2005. *Moosewood Restaurant: Simple Suppers*. New York, NY: Clarkson Potter/Publishers.

______. 2009. *Moosewood Restaurant Cooking for Health*. New York, NY: Simon & Schuster.

______. 2013. *Moosewood Restaurant Favorites*. New York, NY: St Martin's Griffin.

______. 2017. *The Moosewood Restaurant Table*. New York, NY: St Martin's Griffin.

7 "Cruelty-Free" Cruelty

How Vegan Rhetoric Hides the Truth of Banana Production

Lindsay Garcia

Introduction

Like many vegan lifestyle blogs, the website Cruelty-Free Kitty suggests mashed bananas as an ethical alternative to eggs without any investigation into the cruelties of banana cultivation.[1] As a central focus to their imprint, their definition for a "cruelty-free brand" includes that the brands, suppliers, or any third party companies do not test on animals for any ingredient.[2] This website, which highlights mostly cosmetics products, aims to help people who are interested in extending their vegan lifestyle to include non-food related consumables and make appropriate choices that align with their animal-friendly views. While the site helpfully distinguishes the difference and overlap between "cruelty-free" meaning *not tested on animals* and "vegan" meaning *not including any animal-derived ingredients*, these terms often get conflated. In Annie Potts and Philip Armstrong's definition, which implies a coalescing of both, "veganism extends concern for animals beyond a meat-free diet to a purposeful way of life that opposes and avoids the exploitation of animals in any form (including the use of animals for clothing, entertainment, transport, sport, in vivisection, and so on."[3]

Even the Leaping Bunny Program, a certification service which is intended to make consumers feel good about their choices as being devoid of harm to animals, solely focuses on animal testing as a process, rather than all of the potential ways that animals might be exploited in the process of making products. The Program lists in their article "The Corporate Standard of Compassion for Animals" the animals that count the following animals as test subjects: "fish, amphibians, reptiles, birds, and nonhuman mammals."[4] This list does not comprise all nonhuman animals or sentient beings included in the process of making cosmetics or other cruelty-free products, and the statement says nothing about labor practices, environmental impact of chemical production, etc. One genre of animals, in particular, that is left off of this list is pests, especially the insects and nematodes often killed to produce plants. This website and the certification program, as well many similar ones, segregate overall process of production to focus on one aspect, thereby deflecting the full reality of production. They direct the attention

DOI: 10.4324/9781003039013-7

for people aligned with animals to focus on *plant matter* as a way to easily enhance their vegan, cruelty-free lifestyle without highlighting the ways in which the production of plants harms animals, environments, and the human workers who labor within the plant industries. The labeling of different elements of the vegan lifestyle, especially "cruelty-free," serve as a Burkean "terministic screen."[5] Similar to meat being "certified humane" by using best practices to slaughter, certification standards serve as just another term attached to products that does not go far enough toward highlighting all possible cruelties at every stage of operation.[6] "Cruelty-free," as rhetoric, severs the connection between consumers and the material processes of production that might otherwise give consumers the confidence they seek. To exemplify the disconnect between the reality of plant production and the terms used to describe it, I will turn to the banana industry, as bananas are a staple to many vegan diets and lifestyles.

The term "cruelty-free" is said to have derived from the British vegetarian and anti-vivisectionist royal Lady Muriel Dowding who started the company Beauty without Cruelty in 1963.[7] The etymology of cruelty, however, goes all the way back to 13th century Old French meaning "indifference to, or pleasure taken in, the distress or suffering of any sentient being." Beyond the suffering itself, "cruelty" implies a distance from or even a voyeurism of such suffering. "Free" as an adjective, also comes from the same century, though Latin, and is meant to signify a relation close to family that also differentiates between one who is "in bondage" or who is "acting of one's own will."[8] In the phrase, "Cruelty" comes first and is evoked immediately. The distance between production and consumption, in the banana industry and other tropical agricultural industries, can be oceans away—making it easy to be indifferent to what is happening in another area of the world. "Free" is attached as an almost afterthought, drawing up the notion of give-a-ways, lacquered on to hide realities. This phrase thus implies that products that are stamped with cruelty-free approval are without the bondage of indifference or pleasure of suffering—not tested on animals, at the very least. Yet, the phrase actually allows consumers to become indifferent, allowing the label to direct their attention to what is in front of them, rather than following a product back to the seed and the sun. In actuality, the cruelty-free terministic screen masks cruelty in plant production against environments, human workers, and even toward living animals, who do, in fact, test out various aspects of the process. Despite the more visible cruelties happening in the meat industry and through lifestyle products, I contend that vegan studies scholars and vegans themselves need to direct their attention to plant production to see the ways in which animals are harmed sometimes halfway around the world for the plants on the table of everyday Americans.

Cruelty-free labeling closely resembles Marx's notion of commodity fetishism which not only blinds the consumer to the labor that goes into making the object, but also frames the commodity within its value as capital—not as a material object that is living and intermeshed within other species of plant and animal.[9] Even more, eating fruits and vegetables have been touted as potential ways to cure cancer, reverse heart disease, and promote insulin sensitivity, as health cures to fix the effects of capitalism and the comforts of the Global North in human cells.[10] As I show later, this "health" happens at the expense of those who make the food. The mainstream vegan movement's hyper emphasis on vivisection and factory farming, under the guise of offering a "cruelty-free" human existence, elides the fact that significant cruelty has entered fruit, vegetable, and mineral supply chains. As ecologist Daniel Bebber notes, the banana industry "has all of the ingredients of equitability and sustainability issues, disease pressure, and climate change impact all in one."[11]

The scope of examination of cruelty has been highly refracted. While the fields of food studies and the history of capitalism have illuminated the troublesome histories of certain consumer goods, which include slavery, colonialism, and neoliberalism, they have failed to connect habitat fragmentation and environmental racism to such fields.[12] While environmental justice advocates acknowledge pesticides are harmful to farm workers, they often do not consider the nonhumans who are also harmed by those pesticides.[13] While vegan studies, a new subsection of critical animal studies inspired by the vegan turn in the food industry, has interrogated the pitfalls of the meat and cosmetics industries, they have been slow to connect the ways in which plant-based food production harms the environment, other humans, and even animals, and furthermore, how vegan rhetoric further warps such connections. While some recent edited collections have started to interrogate racialization, colonialism, and cultural appropriation within veganist movements, these texts still focus exclusively on the impact of meat-eating and ignore the impact of plant-eating.[14] This chapter aims to draw together these disparate fields and close the gap within vegan studies by emphasizing the contradictions of "cruelty-free" (and related) rhetoric by highlighting various ways that elements of the neoliberal industry of banana production is cruel. In so doing, I argue that the segregated approach to vegan studies, one that emphasizes "meat is bad" arguments hides the ways that environments, laborers, and animals are harmed, maimed, and even murdered in the producing of bananas. I bring an emphasis on plants into the fold of vegan studies and inspire future research to include cruelties associated with plant production practices as part of a cruelty-free lifestyle.

Although bananas have long been seen as a replacement for meat, most people in the Global North, vegan or not, eat or have eaten bananas.[15] The cost of bananas makes them accessible to most people. They are sold in gas stations and often given away for free at supermarkets to children.[16] They are seen as a "healthy" snack, an alternative to junk food, and a staple of the vegan diet. Vegans often tout the banana as a replacement for products whose cultivation involves cruelty to animals, such as nice cream (an ice cream-style smoothie made from frozen bananas) or as a substitute for eggs in baking. North American consumers, in general, do not see banana consumption as an ethical issue, and vegan communities encourage this ignorance. Bananas are largely grown in Central America, South America, and the Caribbean, and thus the success of activism and media attention related to the harms of pesticide use and treatment of migrant workers in California in grape and berry farms has not affected consumers' feelings about bananas. Yet consideration of banana farming's relationship to environmental destruction, inhumane labor conditions (both economically and chemically), the murder of many animals deemed "pests" to protect the plants, and the impact of habitat fragmentation through clearcutting of rainforest ecosystems reveals a troubling picture. The point of this chapter is to demonstrate such cruelties through material examples in order to argue that the directed attention in veganism to eat plants through such terms hides and obfuscates plant production. I start with some background information on how the banana industry grew up around cruelty, bathed in colonial, racist practices, sustains human and nonhuman hierarchies of supremacy. From there, I provide examples that demonstrate cruelties within alleged vegan industries, such as the effects of rainforest clearing, monocropping, and pesticide use, discussing both environmental impact and human suffering. Based on this evidence I argue that vegan studies needs to not take for granted the terms used to describe food and products and instead interrogate terms back to the plants by developing a comprehensive approach to nonhuman species that incorporates plants, minerals, and plant production schema, such as the one I will exemplify here. Since these systems do not take place in a vacuum, one species or classification cannot be segregated from others; these things are part of ecological feedback loops that, in generating human health in the Global North, cause damage to peoples, animals, and environments of the Global South.

Connecting the Rhetoric to the History: An Industry Built on Cruelty

Separating the material conditions of plant production from inequities and cruelties, past and present, is almost impossible; however, the ways in which plants are seen as the perfectly ethical replacement for meat needs to be problematized. Many people from various standpoints argue that eating meat supports the kind of capitalism that exploits animals out of sight of the consumer view, transforming the meat into an *absent referent* for the animal and the conditions they endured during production.[17] Recently, global fast food chains have incorporated vegan options in their menu—a practice elevating vegan choices into mainstream culture and calling attention to the possibility that eating plants also comes from long histories of capitalist exploitation that has affected not only workers, as Marxists and socialists often articulate, but also environments and animals.[18]

United Fruit Company (UFC, later Chiquita Brands International) and Standard Fruit Company (SFC, later Dole Food Company) have held near monopoly status in the U.S. banana marketplace since the early 20th century, having played a major role in the rise in banana consumption in the late 19th century. Founded officially in 1899 by capitalist Minor C. Keith through a series of mergers, UFC began as a series of experiments in banana cultivation designed to cheaply feed the local railroad workers, as the company's predecessor, Atlantic Railway Company, was the major company building railways in Costa Rica. Although President Próspero Fernández Oreamuno of Costa Rica had gifted Keith 800,000 acres of tax-free land and a lease to operate his railroad for 99 years, the railroad business was not profitable enough for his liking. UFC was a merged entity consisting of Keith's Tropical Trading and Transport Company and the Boston Fruit Company, which had been predominately operating in the Caribbean. From there, UFC took over infrastructural and transportation projects in other locations, such as the Guatemalan postal service and the Great White Fleet of steamships, which helped to bring their products to the U.S. markets and control every element of their production. In the early twentieth century, the company UFC merged with over 20 companies, taking over 3.5 million acres of land across Central America and the Caribbean, which gave it ample power to influence governments across all regions where it held land. The company's ability to use its power to control land use, pay little taxes, and continuously expand its market and land power contributed to the coinage of the term "banana republics" by O. Henry, meaning that fruit production and the protection of the capital therein served as an enclave economy that reigned over all areas of politics and life for those who lived in those regions.

Standard Fruit Company, UFC's main banana rival, officially incorporated in 1925 as the result of mergers led by the Vaccaro Brothers. SFC's Central America operations centered in Ecuador. Its domination over the area largely developed much as UFC's developed in Central America and the Caribbean. Since Ecuador's national government did not have the right infrastructure to support such capitalist ventures, SFC teamed up with local governments to build infrastructures and control all levels of production, building box and beer manufacturing plants (to package the fruit and keep the workers happy), thus producing a politically shape-shifting stake in the country. In 1964, SFC merged with Castle and Cooke, which had been largely operating out of Hawaii and contributed to a coup against Queen Liliuokalani in the late nineteenth century, placing the quest for money at the expense of indigenous sovereignty. Dole Food Company was formed, and today it is the largest fruit producer in the world.[19] The corporate colonialism of the banana producers benefited the U.S. economies at the expense of the peoples of the nations where the labor took place.

Cruelty in the form of racial politics was reinforced through corporate strategies of division that pitted the U.S. [immigrant] Black workers against Hispanic and indigenous workers in the Costa Rican banana job market.[20] Racist ideas about Black workers being able to withstand tropical diseases and endure the challenging conditions of plantation labor were transported directly from half a century before when slavery was the primary capitalist mode of producing tropical commodities, like sugar.[21] Language barriers between groups of workers weakened coalition-building and labor organizing, and what labor organizing existed had little power against the power of the industry. In 1968, a UFC merger with capitalist Eli Black's AMK Company grew the company's influence even more.[22] In 1990, the company re-branded as Chiquita Brands International.

Thus, the groundwork of cruelty in the industry was built out of economic desires on the part of large transnational companies and Central/South American/Caribbean governments, many of which were corrupt. By using the banana industry to understand the disconnection between terms like "cruelty-free" and reality of plant production of which goes ignored in vegan studies and the mainstream vegan movements, it is possible to enlarge our understanding of cruelty and build truly cruelty-free movements. The next sections will highlight the environmental, human worker, and animal impacts of the banana industry.

Environmental Impact

The terministic screen that cruelty-free delineates uses environmentally based arguments to advocate for plant-based eating. For example, watering a field of growing plant food takes much less water than raising animals for slaughter.[23] Factory farms likewise have a negative environmental impact; as Jonathan Safran Foer writes, animal excrement from factory farms "seep[s] into rivers, lakes, and oceans—killing wildlife and polluting, air, water, and land in ways that are devastating to human health."[24] As he points out, it isn't that fecal matter in itself is harmful, but the way that the industry overwhelms the production of such. This leads to effect cycles in which overcrowding exacerbates diseases which require antibiotics that then get into the feces and travels through water and airways destroying human and nonhuman health. The added methane in the atmosphere is said to be a major cause of climate change.[25] Thus, plant production is often equated with a well environment and is not seen as potentially damaging to environmental ecosystems. Bananas may, in fact, be less harmful to the local environments where they are produced than factory farming. Because the cruelty conversations around vegan foods are centered around what people are not eating—meat—or the processes that nonhuman animals are being exploited for cosmetics products—vivisection, the academic and mainstream conversations on veganism ignore the fact that plant production processes are still cruel to plants, animals, and ecosystems. Banana farming causes rainforest demolishment, habitat fragmentation, soil depletion (and other impacts) from monocropping, pest control measures, and waste disposal issues, and transporting the crop to markets in other parts of the world has an environmental impact, such as the pollution of the oceans and usage of fossil fuels for powering ships.

The case of Ecuador is illustrative. The country was the world's largest banana exporter in 1954, and it grew bananas on the coastal lowlands. Forests covered around 90% of the country's land. However, banana industry expansion is such that forests only covered about 33% of the region in 1995, and likely constitutes much less of the country today.[26] Deforestation significantly damages biodiversity.[27] While not all of the deforested land went to banana production, a good deal did.[28] In addition to cutting down trees, creating "banana republics" required the deregulation of land use, infrastructures to be developed that furthered environmental impacts, and shifts in government function, which rarely centered principles of democracy.[29]

Establishing banana plants involves capital, labor, and chemicals. Bananas come into being through offshoots from a mother plant. After nine months, the fruit is harvested, and the plant is cut down. This plant has developed a new offshoot, which can be planted to continue the process in cyclical fashion for about ten years, after which the soil has been depleted, so all lingering plant matter is removed by herbicide and cover crops are planted for at least 15 months. While not contributing as many harmful emissions as meat production, the seemingly "natural" process of growing bananas still emits nitrous oxide, carbon dioxide, and methane due to fertilizers and the natural transformation of plants from off-shoot to fruit.[30]

Monocropping has long been the established practice of industrial-scale agriculture for the U.S. growers, since tobacco plantations elevated the U.S. economy such that it became the most powerful in the world. Many side effects of this practice proliferate a number of cruelties. Monocultures produce the conditions for diseases and pests to proliferate, thus requiring hefty amounts of pesticides; they deplete the soil; and they require constant field migration (about every 10 years or less) in order to avoid pests, diseases, and soil, which ultimately leads to more clearing of rainforest lands and the crossing of international boundaries, thus expanding the reach of banana empires across geopolitical borders.[31] Prior to 1950, the main variety of banana was the Gros Michel, a species that was sweeter and stronger than the post-1950 strain that Americans now most widely consume, the Cavendish. In the 1950s, the Gros Michel was decimated by Panama disease, also called "banana wilt," a fungus that lived in the soil, the decimation of which has quantified banana production in the industry from the start. In keeping with a common monocropping practice, the Cavendish became the new sole crop because it was more resistant to Panama disease. However, a new strain of Panama disease, which was found on Columbian farms in 2019 (also called Tropical Race 4), affects the Cavendish and threatens to make bananas completely extinct.[32] Biotech companies have been tasked with developing disease proof plants, which raises the risk of unforeseen issues at the micro- and macroscopic levels, and no feasible alternative to Cavendish bananas has emerged.[33]

Disease has long driven the operations of banana producers. Since the 1940s, UFC has had a close relationship to pesticide companies. Prior to the end of World War II, UFC used copper sulfate (also called Bordeaux mixture) to control Sigatoka, another fungus that infects the leaves of banana plants and limits fruit output. This particular pesticide damages many human (and presumably other mammal) organs, including the skin, stomach, intestines, and respiratory tract. Freshwater fish and invertebrates are heavily affected by its use.[34] To seamlessly connect with agrochemical companies, in 1948, UFC put Dr. Norwood Thorton, formerly of Union Carbide and Carbon Chemicals Corporation, in charge of plant pathology. Around the same time, to control insects, UFC used many different organochlorinated insecticides, such as DDT, Methoxychlor, Chlordane, and Toxaphene, though they found that insects became immune almost instantaneously. This led to the constant rotation of chemicals. Chemical use expanded, and the desire to control pests at every level of production meant that "workers sanitized banana rhizomes before planting them, disinfected their pruning tools, walked the fields on the lookout for disease symptoms, and applied fertilizers, fungicides, herbicides, insecticides, and nematicides."[35] In addition to chemicals, to keep pests at bay, banana plants are wrapped in plastic, which, although recycled, contributes a large amount of waste in areas surrounding banana plantations.[36] The water in residential areas near these plantations is also often contaminated by pesticides, and aerial crop dustings pollute the air.

Infrastructures built up around banana plantations have environmental impacts much as the bananas themselves do. For example, box factories rose up alongside the plantations, which had whole waste infrastructures of their own. New packing factories have been built as recently as 2017. Though framed as employment opportunities for local residents, the waste produced also causes environmental harm, calling up common debates between economists and ecologists that question whether it is more important to provide jobs to impoverished communities even if it puts environments at risk or, alternately, whether the environmental harm and its contribution to climate change threatens the globe so prevalently that no jobs should be provided.[37]

Likewise bananas have a large carbon footprint because of overseas transport.[38] Banana boxes are filled with plastic and other paper products to keep the fruit from bruising as shipments cross the Gulf of Mexico and are fumigated to deter pest and fungal growth during travel. They are placed on palettes and through a series of refrigerated gas-fueled vehicles and ships, they arrive at ports all around the world.[39] Bananas easily rot, so a lot of the work and environmental impact fail to produce edible food. Organic waste is present at every level of production; plants themselves spoil, and bruised bananas are thrown out at washing and packing plants, on transport ships, on trucks, and at the supermarket. A 2018 study shows that bananas are the most wasteful fruit in the U.S. market.[40] This is also because banana peel goes straight to the landfill; in fact, the waste-to-edible ratio by weight is about 2:1.[41] Furthermore, the United States produces more fruit and vegetable waste than meat waste.[42]

While the push to produce organic bananas has improved workers' and environments' health to some extent,[43] significant barriers prevent the scaling of production of organic fruit. For example, organic bananas require 40%–60% more land to produce than conventional bananas.[44] Between complicated legal restrictions and the difficulties growers face upon monocropping without pesticides, these operations remain smaller scale and only account for 9.5% of Chiquita's banana sales and 1% of total organic produce sales.[45] At the same time, organic farming changes the amount of chemicals in the food at every stage of production, but does not reduce them completely as there are still organic compounds that are harmful for humans, nonhumans, and environments. However, the environmental impacts of shipping organic produce and the levels of waste at all stages of production remain cruel and invisible to mainstream vegans and vegan studies due to the rhetoric used to describe how much kinder plants are to consumption than animals. The hidden cruelties, however, extend to labor practices and the workers who engage them for their livelihoods.

Cruelty to Workers

Between commodity fetishism—which separates the product from the mode of production—and the terministic screen which uses "cruelty-free" to direct the attention of consumers to the idea that plants are natural parts of human life rather than colonial capitalist ventures, the farm workers are predominately left out of the rhetoric surrounding plant production. Much of the cruelty toward workers involved in banana production plays out in chemical exposure through water and crop dusting as well as direct exposure at work. In the 1990s, workers won a class-action lawsuit against Standard Fruit for using dibromochloropropane (DBCP, also known as Nemagon), which made them sterile.[46] Although pesticides have been utilized for centuries, following World War II, more industrial chemicals became available on the market in addition to the rise in monocultural farming practices, which made the right conditions for fungi and insects to proliferate.[47] Bordeaux mixture was a common fungicide used to limit the growth of airborne spread of Sigatoka. It caused workers a tuberculosis-like lung infection that was often fatal. What was termed the "white plague" was exacerbated through aerial spraying.[48] In the second half of the twentieth century, many other pesticides were tried as alternatives to Bordeaux mixture, and agricultural scientists treated workers as lab rats.[49] In 1960s Honduras, one way the companies tried to control pests was by limiting the number of subsistence crops surrounding banana plantations, in particular, Chatos. This plantain breed was used to feed banana workers' families and their domestic animals, but banana companies claimed they promoted Moto bacteria. Eliminating these plants, often through coercive means on the side of the companies, significantly damaged food security in the area.[50]

The use of machetes in banana cultivation already poses significant physical danger to workers, but control measures in the 1970s called for dipping the blades into chemical mixtures containing formaldehyde. This caused burning to worker's eyes and potentially permanent damage to fingers, which would incidentally be exposed in the process.[51] In addition to male sterility, DBCP exposure caused workers nausea, vomiting, and diarrhea. In the 1990s, a study of Costa Rican banana workers showed that the rates of leukemia and birth defects were twice the national average; that 20% of male banana workers were sterile. Although these studies contributed to new regulations among banana plantations, enforcement remains uneven.[52] In males, it was common to have melanoma and penile cancer, whereas women were more likely to contract cancer of the cervix.[53] As historian John Soluri points out, the use of chemicals and the lack of care for workers is ultimately about economics; the money made versus the value of a human or nonhuman life and the overall well-being of local environments.[54]

Chemical pesticides reduced the number of workers needed to perform pest control labor by hand as spraying and aerial technologies meant the companies were paying less for laborers. There were fewer jobs and the aerial application increased the health vulnerabilities of the workers.[55] The differing configurations in different national contexts creates labor conditions that vary to some degree, but even where local producers retain some control over their own plantations, multinational corporations govern the shipping process, credit acquisition, and the equipment to maintain production.[56] Challenging conditions for workers began in the very early twentieth century through Minor Keith's building of infrastructural railroads in Costa Rica. Malaria and yellow fever ran rampant through Keith's railway operations, killing an estimated 4,000 workers in the pursuit of these important aspects of the moving of bananas to market.[57]

Potentially adding to the U.S. consumers' willingness to ignore their suffering, most banana workers facing cruelty in banana production are non-White. In the Panama Canal zone of the early 1900s, Black workers were tasked with spraying pesticides to eliminate deadly mosquitoes to allow white people to move into the neighborhood. This task made them vulnerable to tuberculosis and pneumonia.[58] From the beginning, as well, workers would come from all over Central America, the Caribbean, and the southern United States in search of new economic opportunities. In the 1920s, workers from El Salvador, Jamaica, Guatemala, Nicaragua, Belize, Mexico, and other localities in Honduras re-located to the north coast of Honduras to find farm jobs. Because of the volatility and malleability of banana farming, people often moved from job to job. As such, this created both anti-immigrant sentiment and land colonization projects on the north coast, leaving workers open to discrimination by other landowners in the area near the plantations they worked.[59] In Guatemala, Black workers from the West Indies migrated for work and were subjected to the racial attitudes of their White supervisors, most of whom had migrated from the U.S. South.

UFC deployed racial tensions between workers for its own advantage as well. The company systematically created multilingual workplaces by hiring Hispanic workers to keep unions from forming.[60] Labor union solidarity was often broken through racial divides when one group would cross picket lines to obtain the money the other group was striking. For example, in the 1910 UFC Strike in Costa Rica, when the Black West Indies workers refused to show up, the company hired local Hispanic workers and imported almost 700 workers from the British colony of St. Kitts as scabs.[61] When UFC began hiring immigrants in Guatemala in 1923, native workers declined to load a large shipment of bananas onto a steamship in protest and the company hired Black West Indians as scabs. The Guatemalan government initially sided with the Guatemalan workers but switched sides because of its economic ties to UFC and pressure from the U.S. government.[62] This kind of pitting one group against another contributed to a hostile work environment. Such practices have had aftereffects throughout the 20th century that have limited the power of the unions. In one of the most shocking examples of union busting, in 1999 Del Monte corporation kidnapped four union leaders in Honduras, holding 20 members captive in their union hall and demanding they go on the radio and denounce union activities on threat of death. They were only freed when they signed an affidavit denouncing the union, thus forcing them to flee to Guatemala City for two years and later, some took refuge in the United States, for fear for their lives.[63] Death threats were common to labor organizers, and relationships among governments, corporations, and unions remains challenging in Honduras, as human rights violations and abuses remain for people who try to improve the conditions of workers.[64]

Banana growers largely operate in countries where the official minimum wage for an agricultural worker is nowhere close to a living wage.[65] Parents may be forced to bring their children to work with them to increase yields if their employers pay them by their productivity.[66] In 2002, the International Labor Organization (ILO), the U.N. coalition that sets labor standards, estimated that around 69,000 children aged 10–14 were working in Ecuador on banana plantations that year.[67] Though banana workers' unions in Central America have been avenues for women to organize around issues like domestic violence, women typically earn less than men, in part because of the very gendered division of labor in the industry. The majority of women workers work in the packinghouses, where bananas are cut, washed, labeled, and readied for market—working long hours, often seven days a week, though without job security, standing, with repetitive and monotonous tasks wearing sweaty, hot gloves. Single mothers may have to relinquish their children to family members due to the long hours. Chemical exposure and poor working conditions increase the risk of miscarriages and birth defects in their children. The back-breaking work is generally impossible for workers over 40, and employment prospects in other local industries are slim, or no better. What aid is available from NGOs in the Global North may come with assumptions about victimhood that compromise its efficacy. Unions do increasingly make gender equity a priority, but it has not yet been achieved.[68]

Fairtrade bananas, certify that worker conditions meet certain standards, including safety from some chemicals, not using child labor, and better wages, but the percentage of the market remains very small.[69] Although some markets in Europe prioritize fairtrade bananas, the market share in the United States remains below 1%.[70] Major corporations keep banana prices low and their own profits high by ensuring that fairtrade bananas do not go mainstream. Thus, low wages, long hours, child labor use, gendered inequities, union busting, dangerous machinery, and the precarity involved in seasonal migratory work persist. The cruelties of the industry are lurking below terms that equate vegan products with cruelty-free goods. An overemphasis on meat within the vegan industry turns away from cruelties within common plant industries. Even more, many animals are harmed within plant production industries, which gets easily forgotten in meat versus plant debates. The banana industry example will illustrate this phenomenon.

Cruelty to Animals

When the term "Cruelty-free" is thrown around, it implies that no animals are harmed at any stage of the process. However, its precise calculation as being only about animal testing allows people who want to adopt a cruelty-free, vegan lifestyle to ignore cruelty, even toward animals. For example, vegancalculator.com allows a vegan person to enter the amount of years they have identified as vegan and in so doing discern how many animals one has not eaten as well as a number of savings in environmental impact. What's missing from this calculator is how many animals have died in the service of one's vegan lifestyle.[71] Since plants are essential to maintaining a vegan status, a failure to centralize plants and the processes of cultivation in vegan studies or in ethical vegan organizing prevents activist and scholarly endeavors that could help create a more holistic approach to consumption. The truth is that many animals die or become maimed as the result of plant industries, as I will argue with respect to bananas.

First, clearing rainforests is undoubtedly dangerous to the nonhuman animals—as well as other valuable living elements of these ecosystems. Habitat fragmentation, meaning dividing what would be larger habitats into smaller ones to develop pieces of the land for agriculture, reduces species diversity, which is a nice term for *killing animals*. The animals that die may be able to sustain or regrow their populations or they may become endangered because their habitats disappear. Humans think of wild animals in terms of populations instead of individuals, but insects, reptiles, and mammals lose their homes and food sources and die due to lack of resources in places where bananas now grow.[72] Imagine the emotional cost from the standpoint of animal survivors.

The lack of plant and animal diversity caused by monocropping sets the stage for some species to thrive at the expense of others. Panama disease is spread by aphids. Other insects that pose a threat to banana plants include leaf-eating caterpillars, locusts, nematodes, banana weevils/root borers, bees, wasps, fruit flies, and other flying insects, some of which promote another kind of bacteria, Moko. Pesticides cause insect suffering as well as human suffering and environmental impact. Pest animals are rarely even considered within the "animal" category by animal activists. Instead, they are seen as disposable matter without any question of their own lifeworlds—all for the sake of growing international capital markets.

Pest control, however, happens at every level of production, not just in the fields, but in the packing plants, on ships, and in supermarkets at smaller scales to protect the food from insects, rodents, and other small animals. Even beneficial insects, such as ants and lady bugs, are killed by pesticides. Using these chemicals can also produce unwanted side effects, such as the production of new infestations of different kinds of pests, which, in turn, requires new and different chemicals to control, situating the banana industry into a toxic cycle that kills animals and humans who are in close proximity to the spray and adds harmful toxins to food and other products consumed by humans far away. These chemicals easily find their ways into water sources which affect wild nonhumans, such as fish, shrimp, opossums, and skunks, as well as domestic animals like chickens.[73] Studies have also found Costa Rican spectacled caimans and Caribbean coral reefs heavily impacted by such chemicals.[74] By looking at these cruelties within vegan studies frameworks, it becomes clear that segregating animals in hierarchies of cruelty does not eliminate mass animal death, even though consumers are comforted by the terms determinate of their lifestyle.

Can Food Be Cruelty Free?

The popular website Forks Over Knives, one of the leading companies in bringing plant-based eating for health purposes to the mainstream, calls bananas "a superhero of whole-food, plant-based baking."[75] While the consumption of bananas may be healthful, the production of bananas is quite unhealthy for those who plant, cultivate, harvest, pack, and ship them, for the environment, and for animals deemed pests or who live wildly in close proximity to the means of production. The framing of bananas as a "healthy" food highlights the Global North-centric vision of veganism, which places the health of the consumer over that of the producer, and which also engenders a kind of superiority that fails to take into account the plant-based nature of many indigenous diets in the region of production.[76]

My main argument has been that the term "cruelty-free," and by association "vegan," serves as a terministic screen, comforting the consumer, the academic, the vegan into believing a false sense of animal conscientiousness and away from evidenced cruelties at every stage of the production process. In this sense, I used the banana industry, since it has many examples from which to draw on that exemplify deep cruelties throughout history, to the environments local to the farms, to the farm workers, and to the animals on or nearby the site of production. The rhetoric of vegan cultures in the form of "cruelty-free," "organic," and "fairtrade" has presupposed that plant foods and plant cultivation are immune to some of the harms of animal and other agricultural industries. But this is erroneous.

The banana, a healthy staple of a vegan diet, promotes ample forms of cruelty against workers' well-being. To embody the ethics of living a vegan life, one has to contend with the production of plant-based foods. The field of vegan studies must lead the way in developing mainstream public discourses about the cruelties of plant industries to harness the good intentions of vegans and vegan studies scholars to include plants in their research, because not only are cruelties enacted against environments and workers, but animals themselves suffer immensely as the result of this industry. Information is available to consumers, but the labels used to refute the cruelties that exist—fairtrade, organic, cruelty-free—keep consumers from seeing the truths about our food system. Thus, we need to develop a vegan studies approach to cruelties that incorporates plants, minerals, and human impacts with the impacts of nonhuman animals, in order to develop theories and practices that reduce cruelty to the earth as climate change continues. With all of this in mind, I ask, is it possible to be a truly ethical vegan, who engages in a decolonial lifestyle, and advocates for the animals, the workers, and the land? I truly hope so.

Notes

1 Aly Laughlin, "Is There a Such Thing as Cruelty-Free Eggs?" *Cruelty-Free Kitty*, May 22, 2018, https://www.crueltyfreekitty.com/food/cruelty-free-eggs/ (accessed September 16, 2020).
2 Suzana Rose, "List of Officially Cruelty Free Brands 2020," *Cruelty-Free Kitty*, March 23, 2020, https://www.crueltyfreekitty.com/list-of-cruelty-free-brands/ (accessed September 15, 2020).
3 Annie Potts and Philip Armstrong, "Vegan" in Lori Gruen, ed., *Critical Terms for Animal Studies* (Chicago, IL: The University of Chicago Press: 2018): 395.
4 https://www.leapingbunny.org/about/corporate-standard-compassion-animals-standard
5 Burke, "Terminstic Screens."
6 Certified Humane, https://certifiedhumane.org/our-standards/
7 Geri Anne Fennessey, "Color Me Kind," *Vegetarian Times* (June 1997): 60.
8 https://www.etymonline.com/search?q=cruelty+free
9 Karl Marx, "Commodities" in *Capital, Vol. 1: A Critique of Political Economy* (Overland Parks, KS: Digireads.com, 2017): 26–69.

10 Cyrus Khambatta and Robby Barbaro, *Mastering Diabetes: The Revolutionary Method to Reverse Insulin Resistance Permanently in Type 1, Type 1.5, Type 2, Prediabetes, and Gestational Diabetes* (New York, NY: Avery, 2020); Caldwell B. Esselstyn, Jr., *Prevent and Reverse Heart Disease: The Revolutionary, Scientifically Proven, Nutrition-Based Cure* (New York, NY: Avery, 2008); Mayo Clinic Staff, "How Plant-based Food Helps Fight Cancer," *Mayo Clinic*, October 30, 2019, https://www.mayoclinic.org/healthy-lifestyle/nutrition-and-healthy-eating/in-depth/how-plant-based-food-helps-fight-cancer/art-20457590 (accessed September 16, 2020). There are countless stories of people who have cured various ails from eating vegan diets and even more how-to style books which outline the process of doing so. Many evidence-based scientific studies support these claims, the most famous of which is: T. Colin Campbell and Thomas M. Campbell II, *The China Study: The Most Comprehensive Study of Nutrition Ever Conducted and the Startling Implications for Diet, Weight Loss, and Long-Term Health* (Dallas, TX: BenBelia Books, 2016, expanded edition). For a new materialist perspective on the power of food (especially fats) on the human body, see: Jane Bennett, "Edible Matter," in *Vibrant Matter: A Political Ecology of Things* (Chapel Hill, NC: University of North Carolina, 2010).

11 Anna Purna Kambhampaty, "What We Can Learn from the Near-Death of the Banana," *Time*, November 18, 2019, https://time.com/5730790/banana-panama-disease/ (accessed September 17, 2020).

12 Some of the canonic scholarship that addresses cruelties in the banana industry without considering it from a vegan studies perspective are: John Vandermeer and Ivette Perfecto, *Breakfast of Diversity: The Political Ecology of Rain Forest Destruction* (Oakland, CA: Food First Books, 2013); Jason M. Colby, *The Business of Empire: United Fruit, Race, and U.S. Expansion in Central America* (Ithaca, NY: Cornell University Press, 2011); Steve Striffler, *In the Shadows of State and Capital: The United Fruit Company, Popular Struggle, and Agrarian Restructuring in Ecuador, 1900–1995* (Durham, NC: Duke University Press, 2002); Peter Chapman, *Bananas: How United Fruit Company Shaped the World* (Edinburg, TX: Canongate, 2007); Dan Koeppel, *Banana: The Fate of the Fruit that Changed the World* (New York, NY: Plume, 2008); John Soluri, *Banana Cultures: Agriculture, Consumption, and Environmental Change in Honduras and the United States* (Austin, TX: University of Texas, 2005); Steve Striffler and Mark Moberg, eds., *Banana Wars: Power, Production, and History in the Americas* (Durham, NC: Duke University Press, 2003).

13 For example, in these articles, the emphasis is environmental justice and human global health, framed anthropocentrically, with little consideration for the nonhumans affected by pesticide use: Ben W. Brisbois, Jerry M. Spiegel, and Leila Harris, "Healthy, environment and colonial legacies: Situating the science of pesticides, bananas, and bodies in Ecuador," *Social Science & Medicine*, 239 (2019): 112529; Catharina Wesseling, Marianela Corriols, and Viria Bravo, "Acute pesticide poisoning and pesticide registration in Central America," *Toxicology and Applied Pharmacology*, Vol. 207, No. 2 (September 2005): 697–705.

14 Some books that are finally interrogating the whiteness of mainstream veganism and vegan studies, which have come out of non-mainstream animal-focused publishing houses, are: A. Breeze Harper, ed., *Sistah Vegan: Black Female Vegans Speak on Food, Identity, Health, and Society* (New York, NY: Lantern Books, 2010); Aph Ko and Syl Ko, *Aphro-ism: Essays on Pop Culture, Feminism, and Black Veganism from Two Sisters* (New York, NY: Lantern Books, 2017); Julia Feliz Brueck, ed., *Veganism in an Oppressive World: A Vegans of Color Community Project* (Sanctuary Publishers, 2017); Julia Feliz

Brueck, ed., *Veganism of Color: Decentering Whiteness in Human and Nonhuman Liberation* (Sanctuary Publishers, 2019). The following book comes out of an academic press, and out of the whole book, only has three essays that resist the white-centricity of mainstream veganism and vegan studies (by Caitlin E. Stobie, Margarita Carretero-González, and Chanti Chu): Laura Wright, ed., *Through A Vegan Studies Lens: Textual Ethics and Lived Activism* (Reno: University of Nevada Press, 2019.

15 USDA Economic Research Service, "Food Availability and Consumption," last updated February 28, 2020, https://www.ers.usda.gov/data-products/ag-and-food-statistics-charting-the-essentials/food-availability-and-consumption/ (accessed September 29, 2020).

16 Although many supermarket chains include some version of free fruit program for children, here is one example of this: "Giant Introduces Free Fruit Program for Kids," *Supermarket News*, September 8, 2016, https://www.supermarketnews.com/produce-floral/giant-introduces-free-fruit-program-kids (accessed September 29, 2020).

17 Commentators have argued from many different angles that meat production and meat eating supports capitalism. See Josh Berson, "Meat Enables a Capitalism Defined by Inequality," *Salon*, October 20, 2019, https://www.salon.com/2019/10/20/meat-enables-a-capitalism-defined-by-inequality/ (accessed September 17, 2020); Laura Duggan, "Kill Capitalism—Not Animals," *Socialist Voice*, August 5, 2019, https://socialistvoice.ie/2019/08/kill-capitalism-not-animals/ (accessed September 17, 2020); Peter Gelderloos, "Veganism Is a Consumer Activity," *The Anarchist Library*, 2008, https://theanarchistlibrary.org/library/peter-gelderloos-veganism-is-a-consumer-activity (accessed September 17, 2020). Carol J. Adams, *The Sexual Politics of Meat* (London: Continuum, 1990).

18 "Vegan Menu Items at Dunkin Donuts," Dunkin Donuts, June 7, 2019, https://news.dunkindonuts.com/blog/vegan-menu-dunkin (accessed September 17, 2020); Emma Baker, "15 Vegan Gast-Food Items You've Got to Try Now," *Cosmopolitan*, November 22, 2019, https://www.cosmopolitan.com/food-cocktails/a29777119/vegan-fast-food-items/ (accessed September 17, 2020).

19 The coup is missing from Dole Fruit Company's own historical timeline: https://www.dole.com/AboutDole (accessed September 17, 2020), but *Newsweek* reported on it: Krisin Hugo, "How Hawaii Lost Its Last Queen, Liliuokalani, to the Fruit Tycoons at Dole," *Newsweek*, November 10, 2017, https://www.newsweek.com/how-hawaii-lost-its-last-queen-liliuokalani-fruit-tycoons-dole-708512 (accessed September 17, 2020).

20 I am intentionally using the term "Hispanic" here to refer to people who speak Spanish versus the African American workers who speak English. The term is not intended to specifically engender them with ethnic affiliation.

21 Colby, *The Business of Empire*: 3–6. For a history of the rise of the sugar industry, see Sidney Mintz, *Sweetness and Power: The Place of Sugar in Modern* History (New York, NY: Penguin Books, 1986).

22 Black suicided in 1974 after the company sustain major losses as the result of Hurricane Fifi, which toppled Honduran plantations and because he knew that the U.S. Securities and Exchange Commission Was planning to expose the company's corruption, which implicated the Honduran President Oswaldo López Arellano as well.

23 Water Footprint Network, "Water footprint of crop and animal products: a comparison," https://waterfootprint.org/en/water-footprint/product-water-footprint/water-footprint-crop-and-animal-products/ (accessed September 17, 2020).

24 Jonathan Safran Foer, *Eating Animals* (New York, NY: Little, Brown and Company, 2009): 174.

25 Gowri Koneswaran and Danielle Nierenberg, "Global Farm Animal Production and Global Warming: Impacting and Mitigating Climate Change," *Environmental Health Perspectives*, Vol. 116, No. 5 (May 2008): 578–582.
26 Sven Wunder, "Ecuador Goes Bananas: Incremental Technological Change and Forest Loss," in A. Angelsena and D. Kaimowitz, *Agricultural Technologies and Tropical Deforestation* (Oxon, UK: CABI Publishing in association with Center for International Forestry Research, 2001): 167.
27 For a detailed account of the decreasing biodiversity, see: Vandermeer and Perfecto, *Breakfast of Biodiversity*.
28 Wunder, "Ecuador Goes Bananas," 167.
29 Wunder, "Ecuador Goes Bananas," 168.
30 Erik Svanes and Anna K.S. Aronsson, "Carbon footprint of a Cavendish banana supply chain," *The International Journal of Life Cycle Assessment*, Vol. 18 (2013): 1450–1464.
31 Striffler, *In the Shadows*, 31–33.
32 Kambhampaty, "What can we learn."
33 For more information on technological developments, see Koeppel, *Banana*, 185–241.
34 National Pesticide Information Center, "Copper Sulfate," http://npic.orst.edu/ingred/cuso4.html#:~:text=Copper%20sulfate%20is%20a%20naturally,soil%2C%20food%2C%20and%20water.&text=Too%20much%20copper%20sulfate%20can,and%20irritate%20the%20respiratory%20tract. (accessed September 30, 2020).
35 Soluri, *Banana Cultures*, 195.
36 Svanes and Aronsson, "Carbon footprint," 1450–1464.
37 "Columbia: New corrugated cardboard plant for the export of bananas," *Fresh Plaza*, December 15, 2017m https://www.freshplaza.com/article/2186663/colombia-new-corrugated-cardboard-plant-for-the-export-of-bananas/ (accessed September 17, 2020).
38 Svanes and Aronsson, "Carbon footprint," 1450–1464.
39 Svanes and Aronsson, "Carbon footprint," 1450–1464.
40 Elise Kjørstad, "These are the fruits and vegetables we waste the most," *ScienceNordic*, January 4, 2018, https://sciencenordic.com/food-handling-forskningno-society–culture/these-are-the-fruits-and-vegetables-we-waste-the-most/1452912 (accessed September 17, 2020).
41 Ana Bélen Guerrero, Pedro Luis Aguado, Javier Sánchez, and María Dolores Curt, "GIS-Based Assessment of Banana Residual Biomass Potential for Ethanol Production and Power Generation: A Case Study," *Waste and Biomass Valorization*, Vol. 7 (2016): 405–415.
42 Chrstine Costello, Esma Birisci, and Ronald G. McGarvey, "Food waste in campus dining operations: Inventory of pre- and post-consumer mass by food category and estimation of embodied greenhouse gas emissions," *Renewable Agriculture and Food Systems*, Vol. 31, No. 3 (June 2016): 191–201. On the harmful impacts of meat waste, see James McWilliams, "The Deadstock Dilemma: Our Toxic Meat Waste," *The Atlantic*, August 11, 2010, https://www.theatlantic.com/health/archive/2010/08/the-deadstock-dilemma-our-toxic-meat-waste/61191/ (accessed September 16, 2020).
43 Tom Leavitt, "Behind the label: how fair are organic and Fairtrade bananas?" *Ecologist: The Journal for the Post-Industrial Age*, May 28, 2012, https://theecologist.org/2012/may/28/behind-label-how-fair-are-organic-and-fairtrade-bananas (accessed September 30, 2020).
44 Anthony Stoeckert, "Bananas Are Organics Standout," *Produce Business*, September 1, 2018, https://www.producebusiness.com/bananas-are-organics-standout/ (accessed September 17, 2020).

45 For the complicated rules of the U.S. organic farming and labeling, please visit: Electronic Code of Federal Regulations, Title 7, Subtitle B, Chapter I, Subchapter M, Part 205, updated September 15, 2020, https://www.ecfr.gov/cgi-bin/text-idx?tpl=/ecfrbrowse/Title07/7cfr205_main_02.tpl (accessed September 16, 2020); For the Food and Agriculture Organization of the United Nations World Banana Forum guidelines, please visit: http://www.fao.org/world-banana-forum/projects/good-practices/organic-certification/en/ (accessed September 17, 2020); The USDA National Agriculture Library also clarifies what the labels mean when they appear on food items in "Organic Production/Organic Food: Information Access Tools," https://www.nal.usda.gov/afsic/organic-productionorganic-food-information-access-tools (accessed September 17, 2020).

46 Soluri, *Banana Cultures,* 210–214.

47 For a history of the rise of chemical use in agriculture following World War I, see: Edmund Russell, *War and Nature: Fighting Humans and Insects with Chemicals from World War I to Silent Spring* (Cambridge, UK: Cambridge University Press, 2001).

48 Koeppel, *Banana*, 107–109; Soluri, *Banana Cultures*, 198.

49 John Soluri, "La Química," in *Banana Cultures: Agriculture, Consumption, & Environmental Change in Honduras & The United States* (Austin, TX: University of Texas Press, 2005):193–215.

50 Soluri, *Banana Cultures*, 202.

51 Soluri, *Banana Cultures*, 203.

52 Koeppel, *Banana*, 198–199.

53 Catharina Wesseling, Anders Ahlbom, Daniel Antich, Ana Cecilia Rodriguez and Roberto Castro, "Cancer in Banana Plantation Workers in Costa Rica," *International Journal of Epidemiology*, Vol. 25, No. 6 (1996): 1125–1131.

54 Soluri, *Banana Cultures*, 213.

55 Soluri, *Banana Cultures*, 194–199.

56 Striffler, *In the Shadows*, 40–41.

57 Peter Chapman, *Bananas*, 31.

58 Colby, *The Business of Empire*, 88.

59 Soluri, *Banana Cultures*, 76.

60 Colby, *The Business of Empire*, 2–6.

61 Colby, *The Business of Empire*, 102–106.

62 Colby, *The Business of Empire*, 169.

63 Dana Frank, *Bananeras: Women Transforming the Banana Unions of Latin America* (Chicago, IL: Haymarket Books, 2005): 3.

64 Caitrin McKee, "Impunity in Honduras – Will Chiquita Take a Stand?" *International Labor Rights Forum*, September 14, 2015, https://laborrights.org/blog/201509/impunity-honduras-%E2%80%93-will-chiquita-take-stand (accessed September 30, 2020).

65 Working Group 02 Distribution of Value of the World Banana Forum, "Base Study on a Living Wage Methodology for Latin American Banana Production," *Food and Agriculture Organisation of the United Nations* (FAO), http://www.fao.org/fileadmin/templates/banana/documents/WGs_outputs/WBF_WG02_o1_LivingWage_en.pdf (accessed September 17, 2020).

66 Juan Forero, "In Ecuador's Banana Fields, Child Labor is Key to Profits," *The New York Times*, July 13, 2002, https://www.nytimes.com/2002/07/13/world/in-ecuador-s-banana-fields-child-labor-is-key-to-profits.html (accessed September 17, 2020).

67 World Banana Form, "Child Labour in the Banana Industry," Food and Agriculture Organization of the United Nations, http://www.fao.org/world-banana-forum/projects/good-practices/child-labour/en/ (accessed September 30, 2020).

68 Frank, *Bananeras*, 3–6; 13–18.
69 Fairtrade America, "Fairtrade Standards," https://www.fairtradeamerica.org/the-standards/ (accessed September 17, 2020).
70 Aliza Wasserman, "The Other Fair Trade: Why Equitable Banana Matter," Civil Eats, April 7, 2014, https://civileats.com/2014/04/07/the-other-fair-trade-experts-meet-over-equitable-bananas/ (accessed September 17, 2020).
71 Vegan Calculator, https://vegancalculator.com.
72 Nick M. Haddad et al., "Habitat Fragmentation and Its Lasting Impact on Earth's Ecosystems," Science Advances, Vol. 1, No. 2 (March 20, 2015): https://advances.sciencemag.org/content/1/2/e1500052 (accessed September 3, 2020).
73 Soluri, *Banana Cultures*, 197–212.
74 Food Empowerment Project, "Peeling Back the Truth on Bananas," https://foodispower.org/our-food-choices/bananas/ (accessed September 17, 2020).
75 Lisa Esile, "22 Recipes for Ripe and Overripe Bananas," *Forks Over Knives*, April 2, 2020, https://www.forksoverknives.com/recipes/vegan-menus-collections/banana-recipes-ways-to-use-ripe-overripe-bananas/ (accessed September 17, 2020).
76 Julio C. Acosta Navarro, Silia M. Cárdenas Prado, Pedro Acosta Cárdenas, Raul D. Santos, and Bruno Caramelli, "Pre-historic Eating Patterns in Latin America and Protective Effects of Plant-based Diets on Cardiovascular Risk Factors," *Clinics*, Vol. 65, No. 10 (October 2010): 1049–1054.

Bibliography

Chapman, Peter. *Bananas: How United Fruit Company Shaped the World*. Edinburg, TX: Canongate, 2007.

Colby, Jason M. *The Business of Empire: United Fruit, Race, and U.S. Expansion in Central America*. Ithaca, NY: Cornell University Press, 2011.

Frank, Dana. *Bananeras: Women Transforming the Banana Unions of Latin America*. Chicago: Haymarket Books, 2005.

Koeppel, Dan. *Banana: The Fate of the Fruit That Changed the World*. New York, NY: Plume, 2008.

Soluri, John. *Banana Cultures: Agriculture, Consumption, and Environmental Change in Honduras and the United States*. Austin, TX: University of Texas, 2005.

Striffler, Steve. *In the Shadows of State and Capital: The United Fruit Company, Popular Struggle, and Agrarian Restructuring in Ecuador, 1900–1995*. Durham, NC: Duke University Press, 2002.

Striffler, Steve and Mark Moberg, eds., *Banana Wars: Power, Production, and History in the Americas*. Durham, NC: Duke University Press, 2003.

Svanes, Erik and Anna K.S. Aronsson, "Carbon footprint of a Cavendish banana supply chain," *The International Journal of Life Cycle Assessment* vol. 18 (2013): 1450–1464.

Vandermeer, John and Ivette Perfecto, *Breakfast of Diversity: The Political Ecology of Rain Forest Destruction*. Oakland, CA: Food First Books, 2013.

Wunder, Sven. "Ecuador Goes Bananas: Incremental Technological Change and Forest Loss," in A. Angelsena and D. Kaimowitz, *Agricultural Technologies and Tropical Deforestation*. Oxon, UK: CABI Publishing in Association with Center for International Forestry Research, 2001.

8 Maintaining "Meat's" Masculinity

Rhetorical Constructions of Vegan Manhood

Abby Dubisar

Taken together, these cultural sites reveal how—despite its potential to destabilize power structures—vegan discourses rely on the status, power, and profitability of "meat" and masculinity. Further, masculinized terms such as "meat" and "butcher" reflect attempts to reject assumptions that veganism is feminine, since they assure audiences that a person can still perform masculinity with "meat" consumption while simultaneously maintaining an animal-free diet. This trend deserves the attention of rhetoric scholars, especially those interested in feminist interventions at the intersections of power and food, since it potentially masks the critical, political implications of abstaining from eating animals.

Introduction

This chapter interrogates the complex relationship between veganism and masculinity by identifying and analyzing the rhetorical maintenance of that relationship. I use the term masculinity here to represent hegemonic masculinity within a range of masculinities, defining hegemonic masculinity as a masculinity that justifies itself as the dominant, most powerful form of gender expression within a hierarchy that devalues femininity. As Emily Contois writes in *Diners, Dudes, and Diets*, "As a culture's dominant form of masculinity, hegemonic masculinity secures some men's principal position in a society ... it subordinates and marginalizes women and men who are not white, middle class, or heterosexual" (2020, 6). By asking what is more flexible, eating habits or hegemonic masculine gender performance, this chapter examines three themed case studies that interarticulate masculinity with veganism. To that end, the analysis here charts a spectrum or trend that represents the ways in which vegan discourses—vegan butchery, masculine vegan testimonials, and cooking authority through masculine pathways—do maintenance work to attempt to stabilize masculinity, an inherently unstable cultural construct. As an intervention informed by feminist rhetoric, the analysis here works to reveal how veganism potentially destabilizes power structures that gender-based hierarchies then try to fix, an ongoing process.

DOI: 10.4324/9781003039013-8

This chapter joins feminist rhetoric scholarship on food, including taking up the call to address how food-related practices reinforce hierarchies in their representations (Goldthwaite 2017, 7). My analysis also mobilizes Jacqueline Jones Royster and Gesa Kirsch's understanding of the nature of rhetoric as an "embodied social praxis" (2012, 132), as vegan discourses are articulated in embodied ways that rely on social understandings of eating and food, actions with cultural meanings. Because veganism is embodied in gendered ways that potentially destabilize norms, vegans are tasked with locating their veganism on a spectrum related to aligning with or subverting gendered expectations affiliated with cultural meanings of food. Such plotting gets articulated in embodied eating habits and explanations of food practices, both replete with social meanings and values that recur in ongoing processes. The resulting persuasive artifacts and narratives analyzed here represent the complex connections between gender and veganism and the rhetorical constructions that mobilize these connections.

As the chapters in this collection show, veganism holds the potential to disrupt oppressive food systems and promote animal rights. Yet veganism is also increasingly depoliticized and made palatable by growing reliance on white masculinity. While feminist theories detailed in A. Breeze Harper's multifaceted Sistah Vegan Project (2010) and Carol J. Adams's landmark book *Sexual Politics of Meat* (1990) offer ways to disrupt white, patriarchal power structures related to meat-eating, assumed to be the cultural norm, their theories can also be applied to vegan discourses and artifacts themselves. These representations of veganism have the potential to promote white, heteronormative, masculinist tropes in order to create an increasingly persuasive version of veganism that is depoliticized away from its environmental and justice-centered orientations. The recent vegan artifacts analyzed here animate the status quo of white, masculine power in a new context. While my analysis is not meant to promote a singular way to be vegan or even promote animal-free eating, it intends to prompt all eaters to both pay attention to the ways veganism is made persuasive and disrupt the ways in which we are all asked to maintain the white, masculine status quo regardless of the food we eat. In what follows I synthesize literature invested in gender and veganism from a range of fields, including rhetorical studies' existing vegan work, to show how scholars have addressed the relationship between veganism and masculinity. I then analyze three contexts of veganism to show how masculinity is deployed to make veganism palatable and non-threatening. These case studies—vegan butchery, vegan testimonials, and gendered pathways to vegan cooking authority—can be located at specific vegan places and within vegan narratives, plotting them on the spectrum of veganism's attempts to maintain masculinity and mainstream veganism via masculine performance.

Mainstreaming Veganism: A Gendered Project

Within rhetorical studies, scholars have recently begun to address the persuasiveness of veganism and vegetarianism related to its mainstream potential, studies all rooted in the political impulses expressed by not eating animals. For example, Patricia Malesh uses the term "recipes for identity" to describe the narratives vegans are asked to tell by non-vegans in order to explain the choice of eating a vegan diet. These narratives do meaning-making work that reveals "possibilities for resistance, defiance, and change" (2009, 132) that are understandable to mainstream audiences. Malesh's work shows how stories are rhetorically powerful cultural mechanisms that can both construct and demolish ideologies (2009, 143). Other rhetoric scholars have analyzed a number of vegan contexts: how food labels and dietary habits come with ideological values and identity markers that attempt to enter the mainstream (Hahn and Bruner 2012) as well as the ambiguous challenge of veganism's ideological motivation (Jorgensen 2015). Rhetoric's investment in food as rhetorical in its meanings (Burke 1950, 173) can motivate expanding studies to include how such a non-normative choice has to be explained, demonstrating why such choices are persuasive to those who make them.

Recently, meat-free eating has garnered attention and gone mainstream via capitalist ventures that create new contexts for vegan foods, a new rhetorical context yet to be addressed. New vegan and vegetarian foods, including burger-based offerings such as the Impossible Burger and Beyond Meat, participate in a new exigency as they are sold at fast food restaurants, sports bars, and large chain grocery stores. The companies that are manufacturing and selling these burgers invite omnivorous consumers to understand these convenience foods as made for them, familiar yet improved. Scholars beyond rhetoric are paying attention to these new foods, their impact, and meaning. Addressing consumer preference and emerging vegan foods in the eating behavior journal *Appetite*, Peter Slade (2018) describes the identity factors that align with meat-eating, noting how meat and masculinity are associated together in order to avoid feminine connotations of vegetarianism (429). Likewise, writing for the animal science journal *Animal Frontiers*, Lisa Keefe (2018) notes that "companies and activists predict that these new products are the beginning of the end of meat consumption as historically harvested from livestock." However, Keefe tempers that prediction by mentioning that countervailing social, economic, and environmental factors may compromise such a promise, including the cultural construction of meat as a masculine food (2018, 30). Similarly, feminist theorist Carol J. Adams identifies five challenges facing these new vegan food products. Some of these challenges include pricing, scale, and stigmas around technologically developed foods, as well as the fact that "because of the gendered nature of meat consumption, marketing plant-based meat to men poses a challenge" (2018, 126).

Other feminist scholars address broader contexts to show how veganism is interarticulated with gender, with special attention to masculinity and its queer potentials, as veganism is increasingly mainstreamed. Specifically, an emerging body of scholarship questions veganism's political impact related to gender, an effort I join with my study. These scholars analyze a variety of online contexts: vegan food blogs that feature performances of hegemonic femininity and representations of masculinity (Hart 2018) and vegan video blogs that align femininity with control of what enters the body and masculinity with fitness and building muscle, understanding the vegan subject as autonomous, self-reliant, and self-governing and making individuals solely responsible for their own health and well-being (Pirani and Fegitz 2019, 60). Scholars also interview vegans to learn how they explain veganism in relation to hegemonic, patriarchal masculinity, with findings that show how vegan men disagree with the limited definition of hegemonic masculinity, but also do not challenge gender inequality (Greenebaum and Dexter 2018). Scholars also conduct surveys to measure how vegans express identities in relation to race, gender, sexuality, and political beliefs in the context of Trumpism, yielding results that show evidence that vegans' affinity with environmentalists and feminists remains strong, which may hearten those who see veganism as a progressive political practice (Wrenn 2017). Queer potentials for veganism are also taken up by scholarship that questions the binary that justifies veganism as "ethical" eating, placing it in opposition to unethical eating (Overend 2019), identifying the limits of such a dichotomy through a detailed case study of soy, a vegan staple caught up in oppressive systems. Overend's (2019) conclusions resonate with the impulse to trouble the mainstreaming of veganism by arguing that we need to critique food's complicit relationships with dominant power structures (95). Joining in this work, feminist rhetoric's commitment to analyzing and dismantling the gendered hierarchies embedded in food, as well as its understanding of rhetoric as an embodied social praxis, well-position its approaches to join in such interruption of dominant power structures, especially identifying how they continue to be built into food and its meanings.

The time is right for rhetorical scholars to further expand investigations into how gender is expressed through eating and how that identity expression and construction is always unstable. Applying a critical, feminist lens when studying veganism and vegetarianism as represented in artifacts helps us understand how vegan products and vegans themselves mobilize their eating identities by performing gender. Informed by the concerns of food science experts and vegan-feminist theorists, rhetoric scholars can ask: do vegan discourses represent eaters as more able to adjust their expression of masculinity through food or more able to revise their eating habits, regardless of cultural messaging around particular foods?

Case One: Butchering Vegetables

I first take up the term "butcher" as applied to vegan food producers. Attention to this term highlights the instability of vocabulary related to eating animals. At its origins, butcher as a verb refers to slaughter as an action, as a noun as the person who does the slaughtering by dismembering animal bodies into separate parts, or as a noun as a type of shop where animal flesh is sold (Oxford English Dictionary 2020, n.p.). The term is rooted in violence and destructive action, traits of hegemonic masculinity, as well as the impact of ending a life in order to eat. Seeking out vegans' uses of this term shows a reliance on instability and tension, inviting consumers to imagine the violence of animal-based butchery and applying this drama to vegan foods. If vegan foods are assumed to be bland, boring, unsatisfying, and feminine, then "butchers" assert themselves as aggressive and self-assured food producers. Specifically, the Herbivorous Butcher centers this tension in their name. Sibling owners Aubry and Kale Walch (Herbivorous Butcher n.d., n.p.) feature of photo of themselves on their website above a description of their business. In the photo they both wield large knives and gaze directly in the camera in a playful and confident way. The sign on their building is in the shape of a large butcher's knife, evoking the iconography of traditional animal-based butchery in this new context they have created. In their description of who they are, they designate their products as "deliciously deceiving meat-free meats" (Herbivorous Butcher n.d.). The phrase meat-free meats highlights the core theme of instability and the contradiction embedded in this marketing strategy of vegan foods and a physical place to buy them. With rhetoric's attention to the hierarchies embedded in food, as Goldthwaite (2017) invites us to notice and analyze, the Walches rely on the status of butchers and butchery as valued tradespeople and a specialized craft with status in a hierarchy of food workers, due to its masculine identity. They portray themselves as butchers with their knives to adopt this persona that comes with power and nostalgia for a specific version of masculinity in food production. Further, their origin story written for their website details how Aubry Walch (Herbivorous Butcher n.d.) adapted the cuisine of the place where she grew up, Guam, to incorporate familiar flavors in vegan applications. Narratives related to Guam's cuisine or Walch's childhood, however, do not appear in other locations on the website.

By naming their shop a butcher shop, and by putting the term herbivorous first, they pair the unfamiliar with the familiar to create a new context and evoke the nostalgia of a type of store that frequently no longer exists in communities. They assert that their shop is a revision of this lost tradition as well as an homage to that tradition. Its look relies on all the normative expectations for a butcher shop: large glass case with sausages, sliced meats, and whole loaf-shaped meats on display awaiting slicing; butchers in white aprons to answer questions and slice meats; and metal meat slicers to cut these meats to order for customers. Because the shop also sells sandwiches that are prepared to order, some customers might think of their shop as a deli, but the Walches have presented their business as a butcher shop to create a context and assert the unique twist of meat-free meats. Their appeal of ironic nostalgia invites people to look at the visual display of their products and seek comfort in the communal feeling of a neighborhood, accessing new meaning within the familiarity of a context evoking animal butchery.

Mainstreaming veganism by appealing to individuals who do not seek out vegan foods is at the core of this business since the design and theme of their shop aligns with expectations for traditional, animal-based butchery (minus the blood, bones, or other remnants of animal carcasses). This appeal relies on cultural tropes of masculinity that make veganism normative, hip, and steeped in nostalgia as part of a visibly recognizable butcher shop. The styling of the space creates an invitational context that might otherwise be threatening, situating masculine butchery as style to attract those who find a butcher shop comforting, familiar, and appetizing. Their butchery-based strategy is the means to the end of selling vegan meats as a mainstream food that is not threatening to those who eat animal-based protein. For example, when the shop was featured in Guy Fieri's popular Food Network TV show "Diners, Drive-Ins, and Dives," the hierarchy that puts white hegemonic masculinity at the top was on full display, not only through host Guy Fieri's approval of the food, as he declares it delicious after tasting the vegan cold cuts on two sandwiches, but he also brought with him a friend who tried the food and legitimized it as delicious (Daily Motion 2016). In the segment that takes place in the Herbivorous Butcher, the show follows its conventions of a cooking demonstration lead by the owner, in this case Kale Walch, and then testimonials from satisfied customers.[1] White masculine-appearing people, including Fieri's friend, take up the most speaking time as they praise the food. In Fieri's interactions with Kale Walch, Fieri stands in for the "every-man" masculine eater, as he uses words such as familiarity, meaty texture, and other descriptors that speak to the deliciousness of the vegan meat. This clip serves as a persuasive example of the success of the delicious deception described on the shop's website (Daily Motion 2016). They have maintained meat's masculinity through the stylization of their locale, which acts as a visual and spatial invitation to their unfamiliar, potentially alienating food.

In 2017, the Walches tried to trademark the phrase "the vegan butcher" as well as the terms "meat-free meats," "sister butcher," and "brother butcher." They were denied trademark of "the vegan butcher," but granted the others (Crosby 2019). However, shortly after they were denied trademarking "vegan butcher," the company that owns vegan food producer Sweet Earth Foods, Nestlé, was able to trademark the phrase in a different court (Crosby 2019). The Walches are fighting against this discrepancy and questioning why a multinational corporation can trademark the phrase and they cannot, requesting that it remain in the public domain. Sweet Earth uses the term "plant-based" as well as "meat" on their products. In reporting on this case, journalist Jackie Crosby (2019) notes that the Herbivorous Butcher sold $13 million in products in 2018, earnings that reflect the effectiveness of the Walches' strategies and the appeal of their products. Visitors to their shop and website may also note the relatively high sticker price of their product,[2] especially notable for customers who think vegan and vegetarian food should be economical and accessible, again reflecting their location on a hierarchy of vegan foods that places artisanal meat-free meats far above humble vegan fare such as lentils. While the Walches could have chosen a number of styles for their shop, it matters that they chose to create a butcher shop in look and feel, a style that enables them to both subvert and maintain masculine connotations of meat.

Case Two: Vegan Testimonials and Persuasion from the Personal

Vegans themselves rely on masculine tropes to describe and defend their veganism. To see how gender hierarchies are maintained by vegans, reading their descriptions that defend their eating choices shows how they leverage masculinity to maintain its status in a vegan context that could destabilize that status. Robert Atcheson's (2015) description of his own choice to become vegan, entitled "Real Men Don't Eat Meat" and published in *USA Today*, opens with the line, "I fall pretty squarely in the stereotype of an all-American tough guy." By the end of his article, his veganism is fully articulated as a health choice that evolved him out of the so-called standard American diet, but his masculinity remains framed in hegemonic norms that maintain a position of power in a gendered hierarchy. For example, he writes, "As a police captain, I was known as a hard-ass, take-no-excuses kind of guy. Once I went vegan, I gave my officers a message I'd like the whole country to hear: When you eat meat and other animal products, you're playing a losing game of chicken with your health. It takes courage and discipline to ditch that crap and clean up your plate." Thus, Atcheson (2015) reaffirms masculinity's power and positions veganism as a means to access that embodied power. Furthermore, by evading any kind of animal rights or environmental catalyst for veganism, Atcheson (2015) puts masculine human experience at the center of his pro-vegan argument, implying that caring about animals would likely be too feminine for his identity (Birke 1995, 46) and does not align with the courage and discipline he espouses.

Another example is Daniel Kucan's (2013) article in *Huffington Post*, entitled "You Eat Like a Girl: Why the Masculine Dilemma Towards Veganism is No Dilemma at All." Kucan's essay opens with an analogy between eating animals and smoking. Kucan notes that smoking used to be a hallmark of masculine identity and now it is widely known to be toxic, and all the smokers Kucan knows are fraught with regret that they ever started smoking. In telling the story of becoming vegan, Kucan describes adolescent experiences as a dancer, which meant Kucan was often called a "faggot" for choosing to dance. That hobby got Kucan "jacked," however, affirming embodied masculinity in a muscular form, and led to other practices such as Jiu Jitsu and mixed martial arts. The essay intersperses Kucan's preparation for a fight with an opponent called Maldanado, who Kucan describes as a Chinese stylist with Taekwondo agility (Maldanado had defeated Kucan in the past). In describing the scene of the fight, Kucan reflects on the misconception that vegetarians are "soft, effeminate" and builds an argument that eating meat is not a masculine activity. Kucan writes, "You gonna sell me on the hunky masculinity of a huge pot belly, achy joints, wheezy lungs and heart disease? Go try it right now: go spend some time at the Kentucky Fried Chicken and try to pick out one guy, one effing guy, who embodies virility" (2013, n.p.). Kucan argues that virility and strength are qualities of veganism, demonstrated by comments interspersed with his story of fighting Maldanado and knocking him down. Kucan writes that "hunters, predators, bullies, they're all the same" because they think power grants them absolution, but Kucan argues that restraint is the best use of power since humans have dominion over the world but vegans choose to restrain that power (2013, n.p.). Kucan rejects the meat-based masculine identity markers when concluding, "So they can keep their blasé hipster bacon references and their outdoor-cookout meat fests … [they are] too fragile and wavering to resist their own appetites" (2013, n.p.). The essay ends with Kucan noting that a blow to Maldanado lead to Maldanado's tooth becoming embedded in Kucan's elbow, leaving the reader with a vision of Kucan's toughness and intensity since Kucan did not even notice the tooth was there until Maldanado pulls it out. This description aligns with other details Kucan includes, such as describing missing the taste of meat, an experience replicated by Kucan's own bloody teeth and mouth during a fight. Kucan thus maintains masculinity as a way to access power, especially physically, and does not destabilize hegemonic masculine tropes. Instead, Kucan writes how vegans can compete and fight and conquer pressure to express masculinity by eating animals, placing vegan masculinity in a hierarchy above meat-eaters that still maintains a gendered ranking.

A more complex or at least more diverse range of vegan expressions of masculinity can be found in National Public Radio's 2014 series on masculinities, which includes an episode about six vegan men who gather together for a meal for Neda Ulaby's story. The story opens by describing the men as semiprofessional athletes and Ulaby comments that their bodies are "*seriously* built" (2014, n.p.). Ulaby chronicles their descriptions of how their individual masculinity is challenged once people find out they practice veganism. Some of these men address how masculinity is defined and describe ways it can be revised, a move unlike other masculine vegan testimonials that wholly align with masculinity's hegemonic maintenance. For example, Jonathan Katcher states, "Mainstream masculinity is a roadblock to sustainability" (Ulaby 2014, n.p.). In describing the men's philosophies, Ulaby quotes Katcher as insisting that during "an era of climate change and environmental destruction, masculinity should be reframed as protecting the planet, not dominating it" (2014, n.p.). However, the overarching theme of the profile with these men is their personal testimonials that maintain hegemonic masculinity. Also, like the high price tag that accompanies access to the foods at the Herbivorous Butcher, the men in Ulaby's story hold statused jobs, such as Dominic Thompson who "designs very expensive vegan menswear" (2014, n.p.). Thompson does describe, however, the way carework factors into vegan motivations. As Ulaby describes, Thompson grew up in a "rough Chicago housing project," a context in which Thompson saved stray cats and dogs as a response to witnessing people picking on them. Thompson says, "There's nothing more cowardly to me taking advantage of something that's defenseless." So his carework is framed in masculine maintenance, resistance to cowardice.

Beyond the six vegan men gathered at the meal that Ulaby attends, Ulaby's story also features the perspective of academic Kyla Wazana Tompkins who studies gender and food and Tompkins states that veganism is especially masculine because of its emphasis on control. Further, Ulaby includes the views of punk musician John Joseph, author of *Meat is for Pussies: A How-To Guide for Dudes Who Want to Get Fit, Kick Ass, and Take Names* (2014). Joseph uses a tough, intense writing style in his book to address readers directly and issue a wake-up call. It opens by asking, "Who propagated this bullshit that meat makes you macho? My guess is it's the same big business assholes who told you the Marlboro Man was a stud. Well, eating defenseless animals doesn't make you tough, numb-nuts, it makes you a coward" (Joseph 2014, xix). Clearly Ulaby and her interviewees and secondary experts invest themselves in various locations on the continuum of veganism's intersections with masculinity, a moment of progress that somewhat destabilizes traditional patriarchal masculinity as a bastion of power, strength, and domination. Ulaby's (2014) story ends with a conversation with Dan Strong, the person grilling the beet burgers for the meal. Strong is a former butcher and Ulaby wraps up the story with Strong puncturing a masculine myth yet also affirming masculinity. Strong states, "There is an illusion that manhood is like this confidence that is exuded all times. Veganism is that kind of confidence, it really is. It's a choice that we make consciously that we live actively that guides us in our lives. I can't think of anything really more manly than that" (Ulaby 2014, n.p.). With this description, Strong maintains masculinity as firm and stable, even as an illusory cultural construction, a guiding part of his identity. He adapts what he eats to affirm and maintain it.

Case Three: Masculine Pathways to Vegan Cooking Authority

Vegan food companies that sell their products in grocery stores are aware of gendered stigmas around veganism and must mainstream their products in order for them to sell to a broad swath of consumers. The appeals made to market their wares map masculine pathways to create familiar terrain for potential customers to follow. For example, on the 2020 version of the "About Us" section of their website, vegan food producer Field Roast Grain Meat Company describes their process as "rooted in traditional craftsmanship" (n.p.), using vernacular related to artisanal production that aligns well with the hip masculinity purveyed by the Herbivorous Butcher. Founded in 1997 in Seattle and sold to Maple Leaf Foods in Canada for $120 million in 2017, Field Roast uses the word "craft" repeatedly to signify the care and expertise of their products and evoke a sense of nostalgia for non-industrialized foods. They state that they "Believ[e] that tradition and detail matter [so] our artisans draw on time-honored charcuterie methods to carefully craft each of our meats and cheeses" (Field Roast n.d., n.p.). Customers are invited to imagine these unnamed artisans as specialists trained in a long lineage of vegan meat-making traditions and the use of the term "charcuterie," a French term that means "cooked flesh," evokes an elevated, "old Europe" connotation, a backdrop of white masculinity. They also use the term "bold" repeatedly in the description of their products in order to assuage any concern that their foods are weak, bland, or boring. Their descriptions create invitations for customers to not be threatened by their unfamiliar foods.

In 2013, Field Roast sponsored a recipe contest. Recipe contests are meant to elicit excitement over the prizes they can yield and, when sponsored by food companies, create an archive of recipes centered on products the company makes. They also evoke a community feeling around a consumer relationship. On their blog where they announced the contest, they wrote:

> We love hearing from Field Roast fans about the delicious and innovative recipes created with our products. So, we are all very excited to announce our first recipe contest: "Cooks in the Field." We are looking for three people in the community to develop new recipes using Field Roast products. The recipes will be showcased on our website and social media channels for the vegan world to enjoy. We provide the products, you provide the creativity!
>
> (Field Roast 2013a, n.p.)

In order to enter, contestants had to select from one of three gender-stereotyped avatars: two masculine-appearing figures, "grill master" and "urban foodie," and a feminine-appearing figure, "healthy homemaker" (Field Roast 2013a, n.p.). These figures are all represented as forward-facing, grey scale headshot icons with outlines or shadows creating the distinguishable features. The masculine figures seem more desirable, hip, of-the-moment compared to adopting the identity of "healthy homemaker," which not only bears the burden of focusing on health (as opposed to tastiness or pizzazz), but is also positioned in a relational and gendered role as a domestic figure. Unlike "grill master," who has "mastered" something as an expert, or the "urban foodie," who seems cosmopolitan, the healthy homemaker is rooted in a domestic space and serves a family.

With these visual and discursive figures, Field Roast is perhaps trying to make veganism accessible to a non-vegan customer base, but they simultaneously reinscribe the hierarchy that raises masculinity, subordinates femininity, and also restricts the feminine figure to a single, relational role. Readers might question why entrants cannot simply submit their recipe. The gray scale avatars used to represent these three types of authority affix visible gender codes and other icons of food authority. The "grill master" is depicted with short hair and holds a spatula (Field Roast 2013a, n.p.). The "urban foodie" is masculine-coded as well, with a mustache and goatee, wearing a beret (an icon that could be the predecessor to Field Roast's more recent use of "charcuterie") (Field Roast 2013a, n.p.). The "healthy homemaker" is feminized, wearing long hair and holding vegetables (Field Roast 2013a, n.p.). Under each avatar is the line "your photo here" (Field Roast 2013a, n.p.) so entrants are prompted to imagine themselves in these roles. While Field Roast devised a way to have three contest winners and differentiated types of vegan cookery (cooking on a grill, elevated "foodie" cooking, and home cooking for a family), they did not need to utilize gendered avatars to do so. Veganism, Field Roast implicitly argues, need not destabilize the heteronormative family or hegemonic femininity and masculinity. A person can still be masculine and vegan with Field Roast's products, especially by accessing a traditionally masculine grill or a cosmopolitan foodie persona. Similarly, Field Roast's depiction of veganism need not trouble traditional femininity or fuel cultural concerns about feminism or about women rejecting the homemaker role Field Roast reiterates.

When it comes to the narrative descriptions of the contest's three options, Field Roast also invokes gender performance as a pathway to authority. As entrants would expect, Grill Master focuses on grilling:

> Grilling and meat go hand in hand whether at a tailgate or on your back deck. Field Roast Sausages and Frankfurters are obvious choices for grilling, but we also want to see what other creative ideas you can come up with using our range of vegetarian grain meats. If you enjoy making hearty, meaty dishes and are skilled at the grill, do not hesitate to apply.
> (Field Roast 2013b, n.p.)

Here, Field Roast focuses their attention on appealing to meat-eaters and sports fans as well as individuals in the privileged position of owning a grill, which necessitates the class markers of having property and space for cooking outdoors. While it makes sense that Field Roast uses this contest to invite consumers into shared authority over cooking Field Roast products and to advertise to demographic groups who may not connect vegan foods with grilling and tailgating, they could more fully emphasize the food itself over gendered stereotypes, since individuals across the gender spectrum use grills and enjoy sports. In an effort to reject myths that vegan meats cannot be included in grilled meals, often seen as celebrations of masculinity or a part of masculine culinary performance, Field Roast felt the need to emphasize masculinity over the food itself.

With their description of "urban foodie," Field Roast defines the identity and positions vegan innovation in cities:

> We are delighted that Field Roast fans come from across the culinary spectrum, including big city foodies. We are seeking the ultimate Urban Foodie to develop recipes using our range of grain meat products. Do you stay on top of food trends such as creative fusion cuisine and molecular gastronomy? Do you savor the art of dining out, but also enjoy playing in your own kitchen? If your answer to either of these questions is "yes", be sure to enter by October 31!
>
> (Field Roast 2013c, n.p.)

Here cooking becomes a means to staying relevant, contemporary, and mobile, defined as an adventure. Field Roast works to insist that veganism can be exciting and that food is creative and trendy instead of boring. Just like conventional, omnivorous foodie discourses, vegans, too, can fetishize food by using words like "play" and "savor." This description also comes with class markers, assuming that entrants have the economic power and mobility to dine in restaurants and follow trends.

The narrative accompanying "Healthy Homemaker," the feminine coded figure, focuses on cooking for others from a home-based, service position:

> Sharing meals binds us together as families and communities. Do you show your love by preparing food for your family? Are you an expert in balancing good taste and health to please picky palates? We want to see your creative and tasty uses of Field Roast products in your home.
>
> (Field Roast 2013d, n.p.)

The assumed "picky palate" suggests a child, and, thus, this category emphasizes Field Roast's interest in appealing to cooks who feed children, do not want to waste food, and also uphold food as healthy and nutritious. The cook imagined here expresses the emotion of love by cooking meals for others and is positioned in a care-giving role. Just as Field Roast is invested in a version of veganism that does not threaten traditional masculinity, it shows here that veganism can be communal, but is also tied to unpaid feminine labor performed for family members. This description comforts entrants with an imagined vegan mom and insists that vegans do not have to be isolated from families and other communities while also maintaining masculinity's status when placed together with Urban Foodie and Grill Master.

Field Roast shifts expectations somewhat by the outcome of its contest, which named two feminine-appearing people and a masculine-appearing person as winners. Joy Martin became the urban foodie, but the other two, Adam Rosen and Sarah Creighton, align their gendered avatars (Field Roast 2013e, n.p.). In their announcement of the winners, Field Roast defined the contest's purpose: "The goal of this contest was to find cooks in the community to help us build the Field Roast recipe library and share their recipe creations" (Field Roast 2013e, n.p.). Each of the winners has their own blog, and Martin's, entitled "Braisen Woman," focuses on vegan recipes that surprise people who expect vegan fare to be boring. With this choice of a winner, Field Roast shows its intentions to reject stereotypes of veganism. Like Martin, they are invested in showcasing the wide variety of delicious foods vegans eat. Invoking gender stereotypes to negate an anti-vegan stereotype, however, just swaps outdated and disempowering myths and to promote a product. Rosen's blog "Hold the Pigskin," focuses on vegan tailgating food and thus aligns with Field Roast's call for adept grill users, sports fans. In the announcement blog post, Creighton is pictured with children, and Creighton's blog, "Veggie Kids," focuses on making veganism accessible to busy families, including resources like cookbooks and an e-course (Field Roast 2013e, n.p.). Thus, overall, Field Roast succeeds in finding the types of cooks it desired in order to have a range of cooking approaches. However, since it dropped the gendered avatars when announcing the winners, Field Roast implicitly acknowledged that gender stereotypes do not need to be linked to food authority and cooking.

In an attempt to position veganism as masculine, Field Roast felt the need to emphasize masculine roles in order to encourage entrance into a recipe contest as an activity that affirms masculinity. While I do not have access to data regarding whether gendered avatars and pathways matter to contest entrants, the framing of this contest offers an invitation for how other vegan companies ask customers to enact gendered tropes. Notably, when Field Roast ran their recipe contest for the second time, in 2015, no gendered avatars appeared. The only avatar was a chef's hat (Field Roast 2015, n.p.). The announcement blog post is much more inclusive and issues a general call for cooks:

> Two passionate people will be chosen to develop a cheesy vegan recipe of the month … To earn the "Cooks In The Field" title, enter your best [and original] vegan recipe using a Field Roast product … Seize the opportunity to increase your exposure, improve your skills and have some fun! Anyone with a passion for vegan cooking should apply whether you are a blogger, a gourmand, or just an enthusiastic home cook. Do you have what it takes?
>
> (Field Roast 2015, n.p.)

While the gendered tropes do not come into play for this new competition, referring to home cooks last, with the qualifier of "just" does deemphasize and potentially devalue home cooking, which one could assume is the core of Field Roast's customer base since its products are sold in grocery stores. Additionally, some readers could also interpret the chef's hat iconography as a masculine symbol, showing how entrenched gendered codes can be with cooking and food.

Reading Field Roast's contest as a gendered case about masculinity and veganism creates a framework for noticing other examples of masculinity mobilizing and mainstreaming veganism. In a related context about recipes and cooking authority, Bryant Terry's response to the 2014 revelation of vegan blog and cookbook *Thug Kitchen's* (TK) authorship offers another frame for thinking about masculine pathways to accommodate vegan cooking authority—a link that is really about gender, class, and race-based power (Terry 2014, n.p.). TK first began in 2012 as a vegan food blog, idiosyncratically combining vulgar slang language that aligned with stereotypical expectations for "young black men living in low-income urban neighborhoods" (Terry 2014, n.p.) with luscious food photos. As an anonymous blog, the combination of photos and narrative provocatively suggested that vegan authority arose from an unlikely place and an unlikely expert since vegan blogs are most often authored by white women (Hart 2018). However, when the authors of TK were revealed in conjunction with the 2014 publication of their cookbook, the fact that they were young privileged white couple from Hollywood, a man and a woman, facilitated discussion about cultural appropriation and the website's racist implications. Food activist Terry, initially a fan of the website and someone who firmly believes in harnessing pop culture to make progressive change and expand accessibility to veganism, critiques the situation as misrepresentation. Beyond his concern with TK's "masking in African-American street vernacular for [white] amusement and profit" and its attempt to break "unwritten codes governing race and language," both of which TK represents, Terry also exposes how the authors do not know or acknowledge that traditional African and African-American diets are mostly vegetarian (2014, n.p.). Terry educates TK's authors and its readers: "When we peel away the negative stereotypes and reductive portrayals of African-American food, we see a diverse and complex culinary tradition with nutrient-rich foods" (2014, n.p.). The discourse around TK's authorship and Terry's response show the power in harnessing Black masculinity, which TK's authors tried to take on (and off) for dramatic, yet inauthentic and disempowering, effect. It made a difference here that TK's writers performed masculinity specifically—just as Field Roast uplifts it—since it helped them gain popularity and visibility while facilitating a white desire to perform Black masculinity for their own benefit.

In 2020, amid mounting pressure about their long, profitable use of digital blackface and racism in food and food media, the TK authors announced that they would change their name. In a KCRW interview, Terry comments on the context and states, "White people so often refus[e] to listen to Black people and other people of color when they say they're offended or are in pain or that they feel marginalized or unseen" (Kleiman 2020, n.p.). Terry also comments on the "very specific pose" that the TK authors took, appropriating Black comedians and rap lyrics (Kleiman 2020, n.p.). That level of specificity parallels the Walches bespoke design of their shop that is not about a general evocation of the past, but a very specific version of white masculinity related to food. The TK context is ever more prescient since it adopts specific racialized artifacts and styles of Black masculinity for the TK creators' own profit as white people, with no acknowledgment of the long racist history of white people making fun of Black people and pretending to be Black. As the KCRW host details, the issue at hand is power and its relationship to money in food and food media (Kleiman 2020, n.p.). Further, it matters that TK performed Black masculinity specifically, distancing themselves from tropes of white people mimicking Black femininity and food as well as harnessing the credentials of Black masculine cool, hipness, and style for their own white masculine power. On a spectrum of masculine pathways to vegan cooking authority, both Field Roast and TK show clumsy and problematic attempts to enact masculinity as a way to gain influence.

Conclusion

Laura Wright argues in *The Vegan Studies Project* that "[Vegan men] negotiate a discursive space in which that decision [to be vegan] either calls into question their masculinity or attempts to conceptualize veganism as an alternative ultramasculine choice" (2015, 123–124). As these cases show, Wright's either/or binary requires more complication, as vegan men articulate their masculinity at a variety of locations on a spectrum. Just as Overend (2019) troubles the binary between ethical and unethical foods, so too can feminist scholars complicate the binary conception of effeminacy or ultramasculinity expressed in vegan contexts. Doing so joins other scholars in interdisciplinary feminist food studies, including Overend, who argues that "Dismantling a dichotomous approach … can work against … polarization" (2019, 96). Such work opens up conversations regarding the complexity of how veganism addresses gender-based anxieties and maintains hegemonic masculinity in a range of ways, making apparent the linkages between food and gender that re-entrench hierarchies. Likewise, identifying and critiquing the narratives that coarticulate masculinity and veganism shows the lasting status of masculine-based power as valued in a gendered food system. Feminist researchers in rhetoric can continue to address how the embodied social praxis of veganism is made persuasive, identifying why and how masculine pathways assuage anxiety and promote cultural palatability.

Notes

1 What is left unstated in this exchange is how Kale Walch both creates the meat from raw ingredients and then also cooks it and assembles the sandwich from the menu. A parallel sequence for a non-vegan restaurant would be if an owner of a sandwich shop that served animal products was shown slaughtering the animal before cooking the meat and creating the sandwich. However, that moment of slaughter is not part of the eating-based, celebratory, fun vibe of Fieri's show that leaves animal slaughter obscured and unstated. The clip invites this reading and the questioning of where meat "comes from" for the vast majority of restaurants Fieri features.

2 Many would note here that animal-based meat is far more "expensive" than the final retail price shows since it uses vast amounts of resources, such as water, land, and fossil fuels not used in such quantities by vegan meat production. These resources all have environmental costs and monetary costs that are subsidized by federal support of animal industries, all issues far beyond the scope of this chapter.

Bibliography

Adams, Carol J. 1990. *The Sexual Politics of Meat: A Feminist-Vegetarian Critical Theory*. New York, NY: Continuum Publishing.

Adams, Carol J. 2018. *Burger*. New York, NY: Bloomsbury Academic.

Atcheson, Robert. 2015. "Real Men Don't Eat Meat." *USA Today*, July 25, 2015. http://www.usatoday.com/story/opinion/2015/07/25/vegan-diet-health-americacolumn/30514159/.

Birke, Lynda. 1995. "Exploring the Boundaries: Feminism, Animals, and Science." In *Animals and Women: Feminist Theoretical Explorations*, edited by Carol J. Adams and Josephine Donovan, 32–54. Durham, NC: Duke University Press.

Burke, Kenneth. 1950. *The Rhetoric of Motives*. Berkeley, CA: University of California Press.

Contois, Emily. 2020. *Diners, Dudes, and Diets: How Gender and Power Collide in Food Media and Culture*. Chapel Hill, NC: The University of North Carolina Press.

Crosby, Jackie. 2019. "The Herbivorous Butcher Takes on Nestlé in David and Goliath Trademark Battle." Accessed October 13, 2020. https://www.startribune.com/the-herbivorous-butcher-takes-on-nestle-in-david-and-goliath-trademark-battle/565020632/?refresh=true.

Daily Motion. 2016. "Guy Fieri Diners, Drive-ins, and Dives: Vegan Butcher Shop on Food Network." Accessed October 13, 2020. https://www.dailymotion.com/video/x4kmxno.

Field Roast. 2013a. "Announcing the "Cooks in the Field" Recipe Contest!" *The Adventures of Field Roast Grain Meat News*, October 1. http://fieldroast.com/announcing-cooks-field-recipe-contest/.

Field Roast. 2013b. "'Cooks in the Field' Recipe Contest: The Grill Master." *The Adventures of Field Roast Grain Meat News*, October 7. http://fieldroast.com/cooks-field-recipe-contest-grill-master/.

Field Roast. 2013c. "'Cooks in the Field' Recipe Contest: The Urban Foodie" *The Adventures of Field Roast Grain Meat News*, October 14. http://fieldroast.com/cooks-field-recipe-contest-urban-foodie/.

Field Roast. 2013d. "'Cooks in the Field' Recipe Contest: The Healthy Homemaker" *The Adventures of Field Roast Grain Meat News*, October 21. http://fieldroast.com/cooks-field-recipe-contest-healthy-homemaker/.

Field Roast. 2013e. "'Cooks in the Field': Introducing our New Guest Cooks!" *The Adventures of Field Roast Grain Meat News*, November 15. http://fieldroast.com/cooks-field-introducing-new-guest-cooks/.

Field Roast. 2015. "Field Roast Recipe Contest: Enter to Win!" *The Adventures of Field Roast Grain Meat News*, February 10. http://fieldroast.com/field-roast-recipe-contest-enter-win/.

Field Roast. n.d. "About Us." Accessed October 20, 2020. http://fieldroast.com/about-us/.

Goldthwaite, Melissa A., ed. 2017. *Food, Feminisms, and Rhetoric*. Carbondale, IL: Southern Illinois University Press.

Greenebaum, Jessica, and Brandon Dexter. 2018. "Vegan Men and Hybrid Masculinity." *Journal of Gender Studies* 27, no. 6: 637–48. https://doi.org/10.1080/09589236.2017.1287064.

Hahn, Laura, and Michael Bruner. 2012. "Politics on Your Plate: Building and Burning Bridges Across Organic, Vegetarian, and Vegan Discourse." In *The Rhetoric of Food: Discourse, Materiality, and Power*, edited by Joshua Frye and Michael Bruner, 42–57. London, UK: Routledge.

Harper, A. Breeze. 2010. *Sistah Vegan: Food, Identity, Health, and Society: Black Female Vegans Speak*. New York, NY: Lantern Books.

Hart, Dana. 2018. "Faux-Meat and Masculinity: The Gendering of Food on Three Vegan Blogs." *Canadian Food Studies/La Revue Canadienne Des études Sur L'alimentation* 5, no. 1: 133–55. https://doi.org/10.15353/cfs-rcea.v5i1.233.

Herbivorous Butcher. n.d. "About Us." Accessed October 20, 2020. https://www.theherbivorousbutcher.com/pages/about-us.

Jorgensen, Beth. 2015. "To Meat or Not to Meat?: An Analysis of On-line Vegetarian Persuasive Rhetoric." *Poroi* 11, no.1: 1–19. https://doi.org/10.13008/2151-2957.1220.

Joseph, John. 2014. *Meat Is for Pussies: A How-To Guide for Dudes Who Want to Get Fit, Kick Ass, and Take Names*. New York, NY: HarperCollins.

Keefe, Lisa. 2018. "#FakeMeat: How Big a Deal Will Animal Meat Analogs Ultimately Be?" *Animal Frontiers* 8, no. 3: 30–37. https://doi.org/10.1093/af/vfy011.

Kleiman, Evan. 2020. "Digital Blackface: Problems in Food Publishing Aren't New." *KCRW*. Accessed October 20, 2020. https://www.kcrw.com/culture/shows/good-food/black-farmers-juneteenth-cooking-solo/thug-kitchen-revisited.

Kucan, Daniel. 2013. "You Eat Like a Girl: Why the Masculine Dilemma Towards Veganism Is No Dilemma at All." Accessed February 15, 2017. https://www.huffpost.com/entry/veganism-and-masculinity-_b_3764555?.

Malesh, Patricia. 2009. "Sharing Our Recipes: Vegan Conversion Narratives as Social Praxis." In *Active Voices: Composing a Rhetoric of Social Movements*, edited by Sharon Stevens and Patricia Malesh, 131–45. New York, NY: SUNY Press.

Overend, Alissa. 2019. "Is Veganism a Queer Food Practice?" In *Feminist Food Studies: Intersectional Perspectives*, edited by Barbara Parker, Jennifer Brady, Elaine Power, and Susan Belyea, 79–102. Toronto: Canadian Scholars.

Oxford English Dictionary. 2020. "Butcher, n." *OED Online*. Oxford University Press. Accessed October 28, 2020. https://www.oed.com/view/Entry/25324?rskey=kc5Wr9&result=1.

Oxford English Dictionary. 2020. "Butcher, v." *OED Online*. Oxford University Press. Accessed October 28, 2020. https://www.oed.com/view/Entry/25325?rskey=NrIqXc&result=2&isAdvanced=false.

Pirani, Daniela, and Ella Fegitz. 2019. "How Veggie Vlogging Looks Like: Intersections of Gender, Race, and Class in Western Mainstream Veganism." In *Feminist Food Studies: Intersectional Perspectives*, edited by Barbara Parker, Jennifer Brady, Elaine Power, and Susan Belyea, 57–78. Toronto: Canadian Scholars.

Royster, Jacqueline Jones and Gesa Kirsch. 2012. *Feminist Rhetorical Practices: New Horizons for Rhetoric, Composition, and Literacy Studies*. Carbondale, IL: Southern Illinois University Press.

Slade, Peter. 2018. "If You Build It, Will They Eat It? Consumer Preferences for Plant-based and Cultured Meat Burgers." *Appetite* 125: 428–37. https://doi.org/10.1016/j.appet.2018.02.030.

Terry, Bryant. 2014. "The Problem with 'Thug' Cuisine." *Cnn.com*, October 10, 2014. https://www.cnn.com/2014/10/10/living/thug-kitchen-controversy-eatocracy/index.html.

Ulaby, Neda. 2014. "For These Vegans, Masculinity Means Protecting the Planet." *National Public Radio: The Salt*, July 21. Accessed July 28, 2014. http://www.npr.org/sections/thesalt/2014/07/21/332329709/for-these-vegans-masculinity-means-protecting-the-planet.

Wrenn, Corey Lee. 2017. "Trump Veganism: A Political Survey of American Vegans in the Era of Identity Politics." *Societies* 7, no. 4: 1–13. https://doi.org/10.3390/soc7040032.

Wright, Laura. 2015. *The Vegan Studies Project: Food, Animals, and Gender in the Age of Terror*. Athens, GA: University of Georgia Press.

9 Bleeding Burgers

Brutality, Masculinity, and the Vegetarian Narrative

Erin Trauth

Introduction: "Too Much Like the Real Thing"

In November 2016, Anji Bee, one half of the *Happy Healthy Vegan*, a popular vegan web site with more than 215,000 subscribers on YouTube, posted a video of her trying a Beyond Burger for the first time. After cooking and assembling the burger with typical real-meat burger dressings stacked high around the patty, she bites into the burger and exclaims: "Am I even vegan still? This is nuts!" Moments later, Anji's partner, Ryan Lum, tries it and says: "Whoa, this is really good. You know, it's not because this reminds me of a meat, real meat burger. It just tastes really good. It's not gross. I mean some people might think the concept is gross. But no it's just really cool. I know it's 100 percent plant-based. I'm not bothered by it at all." The video, titled "Beyond Burger Taste Test: Too Much Like Meat?" has been viewed 73,000 times as of September 2020, and both the content of the video and YouTube user responses to the video bring up many questions about plant-based burgers and "meats" that look and feel like "real" meat.

In response to the Happy Healthy Vegan's video, several YouTube users immediately noted they simply could not consume a product that was too much "like the real thing." While some proponents of a vegan or vegetarian lifestyle might argue that "life like" plant-based meat alternatives are useful in that they might encourage more omnivores to reduce their meat consumption, others seem to be grossed out—or even horrified—at the fact that some of these products are so realistic. In this chapter, I further examine many of these questions about "bleeding" plant-based burgers. I discuss the ways in which companies, as they expand their vegetarian offerings, mimic (or differentiate from) attributes related to taste, look, feel, and cooking and the rhetorical effect of these choices as we think about the way vegetarian narratives are being shaped and presented to mainstream audiences. Further, I examine "bleeding" plant-based proteins in relation to issues of gender and how such products might influence the connection between meat consumption and masculinity. By doing so, I contribute to a growing knowledge base about new developments in plant-based proteins while diving deeper into the rhetorical, cultural, and gendered issues with "fake" meats that so closely mimic "real" meat.

DOI: 10.4324/9781003039013-9

Returning to the Happy Healthy Vegan's Beyond Burger experience, it is useful to examine some of the immediate responses found in the comments from video viewers.

KITTEN SE DÉVOILE WROTE: "Never will I look at these again … too much like meat for me … I'm officially that vegan."

VEGAN DAVID BOX WROTE "looks too much like actual meat for me …. pass me the sweet potatoes …"

PLANET EARTH WROTE: "… it looks exactly like a real burger …"

SUSAN RAYGOZA WROTE: "yea too much like the real thing, I can't go there. #sadface …"

BRVIHU WROTE: "If fake meat products taste too much like the real thing it actually turns me off. I try to keep telling myself it's not real but I just can't do it."

The comments made in response to Happy Healthy Vegan's Beyond Burger taste test are echoed in a 2019 Mashable taste test of the Impossible Burger, in which Chris Taylor writes: "Many life-long vegetarians are so freaked out by the Impossible burger's resemblance to beef that they simply will not eat it. Some have a visceral distaste; others have been so conditioned by years of dry, mealy veggie burgers that the Impossible tends to trigger alarm bells. It's hard to get used to having your plant-based burger and eating it, so to speak."

On the other hand, many respondents to Happy Healthy Vegan's Beyond Burger video noted the reasons why plant-based burger alternatives could be positive for current vegetarians or vegans:

RASH B WROTE: "The beauty of burgers like this is I realized I loved the texture of meat, not necessarily the meat itself. So this is extremely useful for someone like me who loved meat but likes how being vegan made me feel body wise. This stuff keeps me from cheating, which could lead to falling off the wagon all together. So that's why stuff like this is extremely necessary."

WINCRASHER2007 WROTE: "To me a product like this is ideal for those social situations, like a cookout, where you want to go, but you want to fit in with the "normal" eating that is going on, but still stick to your diet. It's no big deal for you to bring something to throw on the grill. It also helps normalize veganism to other people who may think we are freaks otherwise."

RIHANNA YO WROTE: "People are missing the point of this product. It is not for vegans who DON'T like meat, the idea of it, or who don't want to eat it. It is for those who eat meat, and they're given a much healthier, sustainable option. It is for vegans who actually love and miss meat. It is for those that might struggle with veganism and it gives them a much easier way to transition. Or it isn't for vegans at all, it can be for straight up meat eaters that want a healthier option. It isn't for those that believe "if it bleeds, I won't eat it," or "I don't like meat, that's too realistic, I won't eat it." It's not for your demographic at all. However it is a fantastic product that may encourage those who are not vegan, to choose the healthier and happier option for our planet."

Other respondents noted how such products could be useful in swaying meat eaters over to plant-based alternatives:

S P WROTE: "… if real tasting plant burgers will help people transition to Veganism, let 'em."

ZANE WROTE: "Simulated animal flesh isn't for everyone, and there are certainly reasons to question its existence, but it seems to have some significant value in transitioning society away from actual animal flesh. I buy it occasionally out of curiosity, despite being partially grossed out by the experience."

NY4RL4 7H073 WROTE: "… This is mainly focused at helping meat eaters to stop consuming "real" meat …"

Much of the noted "realness" of these plant-based meat alternatives stems from the texture and look of the patties as well as the fact that they "bleed" when cooked. As more people embrace vegetarian and vegan diets and food options worldwide, many questions have generated about plant-based "fake" meats that more closely mimic "real" animal-based protein meat products in terms of taste, look, and feel. Plant-based meat alternatives are not new to the United States food market, but more recent offerings that have exploded in popularity, including the Beyond Burger and Impossible Burger, as well as a few other new burgers like Don Lee Farms, have one major attribute in common: they "bleed" like real meat when cooked. Some see this as a positive attribute—when a veggie burger not only just tastes like a real meat burger but also looks, feels, and cooks like a real meat burger, it's more likely that meat-eaters might adopt the veggie versions; also, for vegans and vegetarians who can stomach the feeling and look of the burger, this factor can make those who might miss real meat from time to time feel as though they are not "missing out" on the burger experience when they reach for a veggie burger. On the other hand, though, for many ethical vegans and vegetarians, these new burgers create rhetorical problems about the nature of veggie burgers, veganism, and vegetarianism, and some even meet these new veggie burgers—which have a wealth of potential to increase vegan and vegetarian populations across the world—with utter disgust. What does it say about veggie burgers when they are purposefully made to "bleed?" What does it mean that these companies, supposedly at the forefront of vegan and vegetarian movements, opt to intentionally add substances that imitate blood—one of the most blatant reminders of the brutality and inhumanity of meat-eating and one of the very core aspects ethical vegans and vegetarians steer clear of in their quest to save non-human animals? In this chapter, I examine several online responses to the different "bleeding" veggie burgers, including Beyond Burger, Impossible Burger, and Don Lee Farms veggie burger, specifically chat and comment forums and responses to YouTube videos. I discuss the ways in which companies, as they expand their vegan and vegetarian offerings, mimic (or differentiate from) attributes related to taste, look, feel, and cooking and the rhetorical impact of these choices as we think about the way vegans and vegetarian diets are being shaped and presented to mainstream audiences.

Bleeding Burgers: Impossible, Beyond, and Don Lee Farms

When we examine the increasing popularity of plant-based burger alternatives, the Beyond Burger, the Impossible Burger, and the Don Lee Farms "Better than Beef" burger are of certain significance. These three burgers are particularly important because they have made their way into the United States food market mainstream: Don Lee Farms is offered at major markets such as Costco, Sam's Club, Kroger, and Publix. The Beyond Burgers and Impossible Burgers are also offered at several "mainstream" U.S. grocery outlets, and, extending them further into the mainstream, they are available at multiple fast-food or fast-casual restaurants, including TGI Fridays, Sheetz, Dunkin', Hardees, and Carl's Jr. and Burger King, White Castle, Red Robin, and Hard Rock Café, among others. Of particular note about the Beyond Burger is that it is the first plant-based burger of its kind to be sold in the meat section in many grocery stores—since it's "the closest thing" to meat, many have highlighted the company's "plans to stock the burger in refrigerated meat cases alongside beef, pork, and chicken" (Bowie 2016). Interestingly, in "Expanding Our Definition of Meat: Changing Perceptions Of Alternative Protein Sources for Potential Benefits," Audrey Velanovich asks: "If this product looks like meat, is sold in the same grocery store location as meat, cooks like meat, smells like meat, tastes like meat, and even has the same texture as meat can it be considered meat? A follow up question would then be: Why not?" (p. 12). The fact that these burgers are the ones that have been particularly embraced, supported, and marketed through mainstream grocery stores and restaurants is no coincidence—it's the "realness" of these burgers that have skyrocketed them to mainstream attention in a way few soy-based veggie patty of vegetarian bygone could ever achieve.

While new vegan and vegetarian offerings sprout into the market most every day, the Beyond Burger, the Impossible Burger, and the Don Lee Farms "Better than Beef" burger in particular go beyond the quinoa, soy, or bean patties of vegetarian past: specifically, they "bleed" like real meat when cooked. The Impossible Burger, for example, combines genetically engineered yeast with soy leghemoglobin, which is found in soybean roots and contains heme, a molecule also found in animal muscles (Impossible

Burger 2019). Beyond Burger uses beets to simulate the bleeding look and feel (Beyond Burger 2019). For the Impossible Burger, the bleeding was "fundamental to the sensory experience of a burger" and developed purposefully into the development of the product (Williams 2018). Beyond Meat's initial charge was to create the "Maillard reaction," the brown caramelization of foods cooked at certain temperatures, and, later, the addition of beetroot primarily helped the Beyond Burger gain a reddish "meat-like" appearance (Williams 2018). Now, though, Beyond Burger fully embraces the "full" burger sensory experience, proclaiming: "Our belief is that the best way to get people to eat less meat is by giving them more of what they love – in our case, juicy delicious burgers, sausages, and more – without so many of the health, sustainability, and animal welfare downsides of animal-based meats" (Beyond Burger 2019). The Don Lee Farms "Better than Beef" burger "... 'bleeds' organic beet juice and sizzles on the grill from organic vegetable-based fats just like raw beef" (Don Lee Farms 2020).

As referenced in this chapter's introduction, some advocates see this "bleeding" as a positive attribute—when a veggie burger not only just tastes like a real meat burger but also looks, feels, and cooks like a real meat burger, it's more likely that meat-eaters might adopt the veggie versions; also, for vegans and vegetarians who can stomach the feeling and look of the burger, this factor can make those who might miss real meat from time to time feel as though they are not "missing out" on the "real" burger experience. On the other hand, as also evidenced in the comments made in response to the Happy Healthy Vegan's Beyond Burger sampling video, some have met these new veggie burgers with criticism. For many already living a vegan or vegetarian lifestyle, these burgers are *too* much like the real meat product they attempt to imitate, particularly, it seems, due to the "bleeding" that occurs.

What does it say to ethical vegetarians when these burgers are made to "bleed" and look just like raw animal protein? What does it mean that these companies, at the forefront of vegan and vegetarian "meat" movements, add substances that imitate blood—one of the most blatant reminders of the brutality and inhumanity of meat-eating and one of the core aspects ethical vegetarians and vegans steer clear of in their quest to save non-human animals? Using a number of public responses to these actions found across media and academic texts, these questions will be explored further in the following sections.

The image of a reddish, "bloody" burger brings up, additionally, questions of gender. Does the presence of a substance that looks like blood make plant-based burgers more acceptable to male audiences? What does it mean that common manufacturer practices to sanitize the appearance—to remove the visual reminders that these products were once living animal beings—are emphasized for real meats but are now being emphasized by plant-based burger patty companies with the purposeful act of adding that look and feel in? Why is some bleeding—for example, the act of female biological menstruation—still a hidden act—but other types of bleeding, as linked to food consumption, highlighted? Other commenters posit that if plant-based proteins that purposefully bleed and mimic the look of animal meat become mainstream among meat-eaters, the long-running cultural connection between meat and masculinity may be emphasized. Since these companies are explicitly underlining that their products not only taste like meat but also look and feel like meat, people trying them may not be prompted to ask some of the core questions about meat-eating habits and their relationship to gender constructs. One way to make these burgers acceptable to mainstream outlets, of course, is to make them closely mimic "real" meat. In the next section, the stances on "bleeding" plant-based meats will be examined from the ethical vegan/vegetarian's lens.

Bleeding Burgers and the Ethical Vegan/Vegetarian

The increasing popularity of plant-based burger alternatives, in part due to their increasing ability to "look and feel" more like "real meat" is particularly interesting because, in many ways, is paradoxical to the narrative running in real meat products, most of which have been increasingly stripped from their former form to actually *remove* reminders that meat is the flesh of formerly sentient animal beings. In "Cows, cells and the performance of making 'meat,'" Alexandra Sexton writes:

> meat has become a product that simply appears in butchers' shops or on supermarket shelves for many modern consumers. Sometimes it may loosely resemble the animal it was once a part of – such as a whole side of pig or a leg of lamb – but even in these cases it is often devoid of the messier components such as blood and organs. The majority of meat products today are now bloodless, skinless, boneless and often diced into abstract shapes – in short, any reference to the former living body is removed. (6)

Most omnivorous consumers, who are at the end point of the animal consumption process while grocery shopping, are often shielded from the fact that the blood found in animal meat products is actual blood from a formerly living animal being. This shield comes from both a literal blood scrubbing process as well as a figurative scrubbing of any reminder of an animal life that may have been torturous. Sexton writes: "in addition to the meat itself becoming sanitized, the imagery and language around it has undergone a similar process. Images and graphic descriptions of animal slaughtering are not common choices for the packaging of meat products; instead they are frequently adorned with bucolic scenes, careful terminology ('free-range', 'organic', 'hand-reared') and references to the 'happiness' of the animals and the care of the farmers" (7). Wendell Berry, in his essay "The Pleasures of Eating," writes of the disconnect most consumers have with the origins of the food sitting before them on the plate: "... one would not know that the various edibles were ever living creatures ... The passive American consumer, sitting down to a meal of pre-prepared or fast food, confronts a platter covered with inert, anonymous substances that have been processed, dyed, breaded, sauced, gravied, ground, pulped, strained, blended, prettified, and sanitized beyond resemblance to any part of any creature that ever lived" (par. 9). Berry calls this disconnect an "exile from biological reality," which is of ultimate advantage to the food industry (and not to the consumer). The food industry, Berry writes, has "good reason" to veil the connection between food and the farming process. "It would not do for the consumer to know that the hamburger she is eating came from a steer who spent much of his life standing deep in his own excrement in a feedlot, helping to pollute the local streams, or that the calf that yielded the veal cutlet on her plate spent its life in a box in which it did not have room to turn around" (par. 9). In Jason Hannan's book, *Meatsplaining: The Animal Agriculture Industry and the Rhetoric of Denial*, David Nibert illustrates the rhetorical underpinnings at work that have constructed a rhetoric of denial from and about the meat industry: "Today, growing numbers of people have instant access to information, it takes considerable ideological maneuvering to maintain the nefarious belief that most nonhuman inhabitants of the planet are merely 'food' and that their oppression is both natural and necessary" (par. 5). Hannan refers to Carol Adams' "absent referent"—that, for many consumers, the animal referent seems to be removed—or denied—when we pick up a package of meat that was formerly a living and breathing animal.

While the removal of most reminders that real meat is the flesh of a formerly living animal being might effectively help to eradicate a feeling of guilt for many omnivorous meat eaters, for those opting for a fully vegan or vegetarian lifestyle for ethical reasons, this sort of sanitization obviously does not work. For those ethical vegetarians and vegans, then, the purposeful addition of something removed from many "real meat" burgers—namely, the look of blood—to plant-based burgers such as the Beyond, Impossible, and Don Lee Farms burgers—is likely to not only remind them of one of the most unsavory aspects of real meat but also perhaps dissuade them from these alternatives altogether. In *Animal (De)liberation: Should the Consumption of Animal Products Be Banned?* Jan Deckers outlines an "Ethics of Qualified Moral Veganism," relaying personal experiences with a variety of animal beings and how they live their lives for eventual slaughter and human consumption. Deckers explains how "human consumption of animal products supports practices that inflict pain and suffering on numerous animals" (2016, p. 52), and "... the moral response to the death of an animal, and particularly to the deaths of those animals who are biologically close to us, should be one of sadness" (2016, p. 104). While the biological differences of human and animal beings are complex and numerous, one palpable bodily similarity between human and the animal beings most frequently slaughtered for human consumption in the United States—specifically chickens, cows, pigs—is blood. If a human being physically hits a chicken, cow or pig, it may squeal or coil away in response, but whether or not the animal being feels pain and suffering as a result might still be up for debate, since the chicken, cow, or pig cannot necessarily verbally confirm that the hit was in fact painful and that they did in fact suffer. If one physically cuts into the flesh of those animal beings most often consumed by humans, however, blood will undeniably emerge, thus connecting us with animal beings in a way debates about pain and suffering cannot always find common ground within. Blood is a universal reminder of life—with deep ties to religion, reproduction, and culture across the globe—and its presence in meat products is an undeniable reminder that we are linked with animal beings and that the ethics of consuming animal beings is extremely complicated. For those choosing to abstain from eating animal meats, blood—or even the look of blood—is too painful or disgusting of a reminder of our intrinsic link with animal beings. When the companies now at the forefront of vegan and vegetarian "plant-based meat" movements with determination add substances that imitate blood—one of the more flagrant reminders of the brutality and inhumanity often associated with meat-eating and one the most profound connections humans have with animal beings often consumed by humans, it's simply too much for many vegans and vegetarians, even if the product is in fact fully animal-free. While the blood found oozing from animal-based meats is obviously—and usually either accepted or ignored—as a byproduct of the death of an animal, the addition of a blood-mimicking element to an otherwise plant-based product, which would not naturally exhibit that element, may signify too much of a connection—too strong of a reminder—that plant-based products are trying to imitate a product many vegans and vegetarians actively shun. Returning to Hannan's rhetoric of denial—those rhetorical turns that have pervaded Western society for decades and convinced so many consumers to engage in "considerable ideological maneuvering" to ignore (or disregard) that the blood found on one's plate is, in fact, the blood of a previously living non-human being—perhaps the only reminder left after the meat has otherwise been scrubbed of bones, eyes, organs, and placed in plastic wrap—the purposeful addition of a blood-like substance by plant-based meat manufacturers hits too close to home. For some, it may sicken or repel them further from these otherwise innocuous meat-free options, leading one to ask: should we just leave the blood-like substances out of the equation, and let plant-based "meat" look and feel like what it is—plants?

For other vegans or vegetarians, the presence of blood might not deter them from trying these plant-based bleeding burgers, and might in fact aid in their ability to stick to their animal-consumption-free lifestyle. J. Kenji López-Alt writes: "But many ethical vegans still enjoy the taste of meat and crave it. Plant-based alternatives are a good way to satisfy those desires while still staying true to basic principles" (2016). YouTube user rash b wrote: "… This stuff keeps me from cheating, which could lead to falling off the wagon all together. So that's why stuff like this is extremely necessary." As additionally evidenced in respondent comments to the Happy Healthy Vegan YouTube video, some proclaimed meatless eaters see social benefits in having a "realistic" plant-based option, as in the comment quoted above by YouTube user wincrasher2007, who believed that such products may help "normalize veganism" as they help vegans participate in food-centric rituals. In these ways, the bleeding aspect of the burgers adds to the sensory and social attributes of the plant-based burger options, allowing vegans and vegetarians who might feel as though they are missing out or are even ostracized for their lifestyle to "stay on the wagon." This also connects to the notion that, somehow, the standard Western diet has normalized that we should see the reddish tinge of blood when we consume proteins; somehow, *not* seeing blood on the plate is abnormal. Interestingly, some believe that plant-based meat products might deter people from sticking with a plant-based lifestyle. In "Say No To Faux," J. Kenji López-Alt writes: "… faux meats remind vegans and vegetarians of what they are missing. There must be a couple of people out there (vegan or otherwise) who truly prefer the flavor of mock meat products over the real deal, but for the most part, the praise you hear for mock meat comes in the form of, "It's *surprisingly* delicious!" (2012). The element of surprise is still a common theme in reviews and media of new plant-based products; many reviewers seem flabbergasted that vegan products—still the abnormal option across most diets—can not only not taste terrible, but that they can also taste good.

Others still feel that plant-based burgers are beneficial, especially when it comes to swaying omnivores. Also referenced in comments made in response to the Happy Healthy Vegan YouTube video is that the addition of burgers that are so "lifelike" to the market is beneficial in the larger picture of animal well-being and liberation because such products will increase omnivores' desire to try meatless products and adopt the consumption of more meatless meals. Indeed, this potential for change is evidenced first by the prominence of options present in conventional grocery stores and fast and fast-casual restaurants. Long gone are the days of merely a salad option for those wanting a vegan or vegetarian meal when eating in mainstream locales. Further, these options have garnered significant media attention—when burgers such as the Impossible and Beyond burgers first debuted, many mainstream media outlets—from *The Washington Post* to *Cooking Light* to *USA Today*—covered the products, offering taste tests with often raving reviews on the ability of plant-based ingredients to combine and make something so lifelike and real. A 2019 Washington Post taste-test article noted the Impossible Burger even had a "steak like ooze" and that it "struck a meat like pose by having a convincingly slightly pink center and browned exterior" (Heil 2019). A 2017 episode of The Today Show featured its anchors biting into Beyond Burgers, with Craig Melvin exclaiming that "it's not just for vegetarians, it's also for folks who are health-conscious."

Dr. Oz featured the Impossible Burger on his popular show, comparing its attributes to a "traditional" beef burger and asking his viewers to share videos of themselves trying out the "meatless meat fast-food craze."

The makers of the referenced plant-based burgers are also owning the idea that these products are so close to animal meat that they can serve as viable options for omnivores and carnivores, too: "Beyond Meat's founder Ethan Brown said the goal is to introduce a healthier, cruelty-free option to omnivorous consumers who aren't venturing to the freezer aisle for their burgers—evidenced by a new slogan on Beyond Meat's website that reads 'Carnivores welcome'" (Bowie 2016). Brown states: "A lot of people love to eat meat. What I'm doing is allowing them to eat more of what they love, more meat. It's just meat from plants. That's the only difference" (Bowie 2016). Brown's statement about the plant-based burgers actually being "meat" from plants brings us back to Velanovich's question reconsidering what we think of as "meat." If meat was to be re-conceptualized, might this draw in even more potential animal-based meat eaters? This, of course would involve debate about who can claim ownership and use to the term "meat," which has been problematized and has already come into question in 2018 when Missouri became the first state to ban the use of the word "meat" on any products except those containing actual animal flesh (Gibson 2018). For now, the underlying question remains for some—what makes meat *meat*?

For those concerned with animal rights and liberation, the definition of meat certainly centers on animals. And those who abstain from the consumption of animal flesh, the presence of blood—even if that blood is in fact derived from plants—is too close for comfort when it comes to buying into newer plant-based burger options that "bleed." While the potential benefits of these products are apparent for groups of vegans or vegetarians who may miss the taste and feel of "real" meat or fear they may miss out on social opportunities if they don't eat something appearing close to "real" meat, blood-like substances are enough to deter others. Undoubtedly, burger options such as the Beyond, Impossible, and Don Lee Farms burger are luring in the omnivorous, opening up the potential for more people to opt into even a partially animal-flesh-free diet. The long-term impact of what it means rhetorically that these burgers must "bleed" to do that luring successfully is another question, though. In the next section, issues related to gender and the plant-based bleeding burger will be introduced.

Gender and the Bloody Burger

For many, the image of a reddish, "bloody" burger may also bring up questions of gender. The act of simulating bleeding is perhaps making the plant-based burgers more acceptable for male consumption; without the bleeding, the burgers might otherwise be seen as too feminine (Juschka 2020). Camilla Mørk Røstvik, who researches the visual impact of blood, questions why blood associated with the death of animals is seen as not only acceptable but as desired (as opposed to depictions of human blood, which is more often hidden purposefully, especially as it relates to menstruation). She writes: "It is strange how bleeding that signifies a healthy human body is invisibilized, but bleeding that is associated with hunting and violence, blood that means you have killed an animal, is acceptable" (Røstvik qtd. in Williams 2018).

Others posit that if plant-based proteins that purposefully bleed and mimic the look of animal meat become mainstream among meat-eaters, the long-running cultural connection between meat and masculinity may be emphasized. Since these companies are explicitly underlining that their products not only taste like meat but also look and feel like meat, people trying them may not be prompted to ask some of the core questions about meat-eating habits and their relationship to gender constructs. One researcher asks: "Are these plant-based meat alternatives sufficiently different that they will challenge existing ideologies?" (Elder qtd. in Todd 2018). Once again, this question brings to mind the overlying inquiry into what makes meat "real"—and what a food must look and feel like for it to be considered a meat product. For others, however, the bleeding might help actually alleviate the gendered issues with the consumption of vegetarian products. In her research on the relationship between men and meat, Emma Roe suggests that the more meatless options that are available in mainstream outlets, the better, as the widespread exposure can lead to more male "social permission" to eat vegan and vegetarian options (Roe 2018). One way to make these burgers acceptable to mainstream outlets, of course, is to make them closely mimic "real" meat, meaning that burgers such as the Beyond, Impossible, and Don Lee Farms burgers can serve as a viable gateway for men to experiment with and enjoy vegan or vegetarian products without fear of social disapproval.

Additionally on the subject of gender, questions about masculinity and even what it means to be a true "American" might come to play. In "Men, Meat, and Hegan Identity: Veganism and the Discourse of Masculinity," Laura Wright discusses the "male activity of eating meat," explaining that the American way of life has so far lionized the consumption of meat that we have also seen a "profound denunciation of vegetarian and vegan diets as indicators of weakness, ethnicity, and femininity, all of which have been constructed as threats to a traditional 'American' way of life" (2015, p. 114). Men, then, are particularly at risk for being considered weak—or perhaps even un-American—if they abstain from consuming animal products. If plant-based burgers "bleed" on the grill and look almost like the real thing, especially when consumed in the presence of other males at backyard barbeques and tailgates, then these products might help those men not feel—or appear—as "weak."

Further, connections between meat consumption as pleasure—principally a male-centric pleasure—are of consideration when discussing the potential appeal of "bleeding" plant-based burgers. In "Framing the Future of Food: The Contested Promises of Alternative Proteins," Sexton et. al write that meat alternative protein companies have been careful to toe the line in being "healthier" but not devoid of pleasure, particularly pleasure as it might be associated with male "macho-ness":

> Central to this balancing act is the work being done by AP (alternative protein) developers to make their products viscerally indistinguishable from their conventional counterparts, and reveals the 'dirty' foundations of their clean-eating promise. Rather than a choice of abstinence and worthiness, APs instead offer a tasty and hedonistic way of saving the world. Making plants bleed and cells sizzle retains a degree of animalness – and arguably macho-ness – which in turn offers the pretense of defilement that is deeply associated with the consumption of animal flesh, as well as other animal-derived products.
>
> (Sexton et al. 2019)

The authors argue that an emphasis on pleasure—specifically male pleasure—is touted on many social media spaces for alternative proteins: "Such framings call to mind recent trends elsewhere that have worked to rebrand plant-based eating as a more hedonistic, visceral experience: these include the launch of vegan fast food chains with provocative names and a recurring visual trope of plants set within a 'bloody' scene being chopped up by a cleaver-wielding chef or butcher (most often a white, hyper-masculine male) (2019). Substances that look like blood, then, oozing from plant-based burger options such as the Beyond, Impossible, and Don Lee Burgers, play into a "promise" of such foods to "look and perform" like animal-based foods help to "appeal to the carnivorous (typically male) consumer, one who has yet been unable to give up the sensory pleasures of animal foods and now does not have to" (Sexton et al. 2019). The links between masculinity and meat consumption illustrated by Sexton et al. have been noted in many other sub contexts (Adams 1990; Freeman and Merskin 2008; Stanescu 2016), but the question as to whether "bleeding" plant-based burger products such as the Beyond, Impossible, and Don Lee Farms burgers will be sufficient to convince omnivorous or carnivorous males to stick with animal-free alternative proteins in the long-term. The popularity and availability of these products in the mainstream is but yet only a few years in the making, and the future of how these products can influence the future of vegan and vegetarian buy-in remains to be seen.

Returning to Nibert's discussion of the "considerable ideological maneuvering" that many consumers may experience, leading to a "rhetoric of denial" (Hannan) about meat and its ties to a real, living being, concepts of gender and its relation to meat consumption are but one layer of untangling that may need to take place when we consider the true impact of blood-like substances being added to otherwise plant-based products. The weight of cultural expectations about what belongs on our plates is not just cultural and ideological but also rhetorical, and every image of plant-based options portrayed in public media matters. If plant-based meat producers such as Beyond, Impossible, and Don Lee Farms continue to add elements that make their plant-based meats look and feel more like animal-based meat, what rhetorical impact will this have on future plant-based brands that seek a purer, bloodless image and feel for its products? Will, for example, a male be as likely to try it, or offer it at a cookout? Will omnivores interested in trying plant-based options be so used to seeing plant-based options that "bleed" that a simple nut-and-vegetable patty, sans any blood-like additions, will seem bland or unappealing? It is these questions about the visual rhetoric of blood elements in plant-based meat options that still linger.

Conclusion: The Future of "Bleeding" Burgers

While the sight of blood-like substances in burgers such as the Beyond, Impossible, and Don Lee Farms burgers are simply too much of a palpable reminder of the brutality toward and death of animals necessary to produce animal meat products, from a global ethical and environmental standpoint, plant-based burgers seem, at the very least, an additional means by which some are able to help maintain a plant-based lifestyle and perhaps a gateway to the conversion of some omnivores and carnivores to a more plant-based lifestyle. Even in light of prevailing attitudes about masculinity and meat, especially in the United States, the promise of the pleasure many derive—or derived in the past—from "real" meats can seemingly be imitated through the addition of elements adding to the look and feel of an otherwise animal-free burger product. These products are of especially important consideration when we consider our current global issues of climate change and possible links between animal agriculture and disease (including pandemics). In "The Ethics and Politics of Plant-based and Cultured Meats," Jeff Sebo writes:

> Recent developments have raised an intriguing possibility: that science, technology, and meat are not only the cause of (or at least a partial cause of) but also the solution to (or at least a partial solution to) the current food crisis. In particular, researchers have made tremendous strides in developing plant-based meat (i.e., meat that comes from plants) as well as cultured meat (i.e., meat that comes from a cell culture). Many organizations are now developing these products, and with each passing year they are reaching new milestones in terms of quality, quantity, and affordability. If this progress continues, then plant-based and cultured meat could be a game changer: unlike other alternatives to our current food system, they could allow people to eat what they want while eliminating many of the human, nonhuman, and environmental costs of what people currently eat.
>
> (2018)

Understanding that animal meat production has created many of the global issues we are now facing head-on, it is interesting to think that a version of "meat"—depending on how we define meat in the future—may, in turn, serve many of the solutions we need to save the globe from ethical, environmental, and public health problems. If the addition of a plant-based blood-like substance is what it might take to have people eating in plant-based ways while still "allow[ing them] to 'eat what they want,'" then burger brands such as the Beyond, Impossible, and Don Lee Farms burgers may indeed be at the forefront of a major movement to bridge together these—for some—formerly mutually exclusive goals. While the addition of "bleeding" burgers adds a layer of complexity to definitions of meat and of plant-based eating as well as challenges many already plant-based eaters, the potential for these products to serve as a gateway food for otherwise omnivorous or carnivorous eaters is a movement that warrants further consideration, especially if we are to utilize plant-based food options as a vehicle to help solve larger global problems in the future.

Bibliography

Adams, Carol J. 1990. *The Sexual Politics of Meat: A Feminist-Vegetarian Critical Theory*. New York, NY: Continuum.

Berry, Wendell. "The Pleasures of Eating." Accessed June 1, 2022. https://www.ecoliteracy.org/article/wendell-berry-pleasures-eating

Beyond Meat. 2019. "Frequently Asked Questions." Accessed September 20, 2020. https://www.beyondmeat.com/faqs/

Bowie, Richard. 2016. "First Vegan Burger to Be Sold Alongside Meat Debuts." Accessed September 21, 2020. https://vegnews.com/2016/5/first-vegan-burger-to-be-sold-alongside-meat-debuts

Deckers, Jan. 2016. "The Ethics of Qualified Moral Veganism." In *Animal (De)liberation: Should the Consumption of Animal Products Be Banned*, 51–106. London, UK: Ubiquity Press.

Don Lee Farms. "Organic Plant-Based Burger." Accessed December 2, 2022. https://www.donleefarms.com/product-page/organic-plant-based-burger.

Freeman, Carrie and Debra Merskin. 2008. "Food for Thought: Essays on Eating and Culture." In *Having it His Way: The Construction of Masculinity in Fast Food TV Advertising*, edited by Lawrence C Rubin, 277–293. Jefferson, NC: McFarland.

Gibson, Kate. 2018. "Missouri First U.S. State to Ban Word "Meat" on Anything but Animal Flesh." Accessed September 22, 2020. https://www.cbsnews.com/news/missouri-first-state-to-ban-word-meat-on-anything-but-animal-flesh/

Hannan, Jason. 2020. *Meatsplaining: The Animal Agriculture Industry and the Rhetoric of Denial*. Sydney, Australia: Sydney University Press.

Heil, Emily. 2019. "Impossible vs. Beyond: We Tested Cook-at-Home Versions to See Who Makes a Better Vegan Burger." Accessed September 20, 2020. https://www.washingtonpost.com/news/voraciously/wp/2019/10/03/impossible-vs-beyond-we-tested-cook-at-home-versions-to-see-who-makes-a-better-vegan-burger/

Impossible. 2019. "Heme and Science." Accessed September 21, 2020. https://impossiblefoods.com/heme/

Juschka, Darlene. 2020. *Contours of the Flesh: The Semiotics of Pain*. Sheffield, YO: Equinox eBooks Publishing.

Kenji Lopez, J. 2012. "Say No to Faux." Accessed September 20, 2020. https://www.seriouseats.com/2012/01/the-vegan-experience-day-5-say-no-to-faux.html

Kenji Lopez, J. 2016. "Let It Bleed (Humanely): We Taste the Vegan 'Impossible' and 'Beyond' Burgers." Accessed September 20, 2020. https://www.seriouseats.com/2016/10/beyond-burger-impossible-burger-vegan-taste-test.html

Merinuk, Madeline. 2019. "Are Impossible Burgers Healthier Than Beef Burgers? Here's the Verdict." Accessed September 20, 2020. https://www.doctoroz.com/article/are-impossible-burgers-healthier-beef-burgers-heres-verdict#comments

Parker, John. 2019. "The Year of the Vegan." *The Economist*, December 2018. https://worldin2019.economist.com/theyearofthevegan

Roe, Emma. 2018. "Sun's Out, Buns Out: Exploring the Alfresco Ritual of Meat, Fire, Man's Work, and Sustainability." Accessed September 20, 2020. https://www.foodsecurity.ac.uk/blog/suns-out-buns-out-exploring-the-alfresco-ritual-of-meat-fire-mans-work-and-sustainability/

Sexton, Alexandra. "Cows, Cells and the Performance of Making 'meat'." Accessed September 20, 2020. http://www.semicopia.com/cow-cells-and-the-performance.html

Stanescu, Vasile. 2016. "The Whopper Virgins: Hamburgers, Gender, and Xenophobia." In *Meat Culture*, 90–108. Netherlands: Brill.

Taylor, Chris. 2019. "Impossible vs. Beyond: Which Tastes More Like Real Beef and Does It Matter?" Accessed September 20, 2020. https://mashable.com/article/beyond-meat-impossible-burger-review/

TODAY. 2017. "Can TODAY Anchors Tell Which of These Burgers Is Meatless?" Accessed September 21, 2020. https://www.today.com/video/can-today-anchors-tell-which-of-these-burgers-is-meatless-981295683811

Todd, Sarah. 2018. "Beyond Meat and Impossible Foods Burgers Could Change the Way We Think About Masculinity." Accessed September 20, 2020. https://qz.com/quartzy/1603993/beyond-meats-vegan-burgers-could-change-the-way-we-think-about-masculinity/

Velanovich, Audrey. 2018. "Expanding Our Definition of Meat: Changing Perceptions of Alternative Protein Sources for Potential Benefits." Accessed September 20, 2020. https://www.theworldmind.org/home/2018/10/19/expanding-our-definition-of-meat-changing-perceptions-of-alternative-protein-sources-for-potential-benefits

Williams, Lara. 2018. "Why Even Vegans Crave Burgers That 'Bleed'." Accessed September 20, 2020. https://www.vice.com/en/article/nepbad/why-even-vegans-crave-burgers-that-bleed

Wright, Laura. 2015. *The Vegan Studies Project: Food, Animals, and Gender in the Age of Terror.* Athens, GA: University of Georgia Press.

10 Chicken Without Chicken, Sausage Without Sausage

Rhetorical Remediations of Vegetarian and Vegan Foods Recalling Meat

Francesco Buscemi

Introduction and Existing Studies

Meatless burgers, turkey-free breasts, salami meat-free sandwiches, "Tofurkey Italian Sausage" (Hackett 2019) and many others constitute an interesting subcategory of vegetarian and vegan foods. Their commonality is the fact that all of them are linked to the animal and its parts by exactly what they do not contain, being vegetarian and vegan foods. Moreover, they are not niche products available in small shops for a narrow public; rather, they are popular brands that may be found on the supermarkets' shelves in both the US and Europe. The fact that these products are available in many countries, which are based on very different approaches to food, makes the issue even more interesting. Thus, this chapter analyzes the communicative strategies adopted by big retailers' websites in three countries (the US, the UK, and Italy) in order to sell vegetarian and vegan products which constantly recall meat in their names. This study thus focuses on the rhetorical and communicative strategies adopted by commercial websites in order to draw the attention of customers.

The rationale for this study starts from the idea of decoding a language that has been subverted to the point of signifying the opposite it usually means. Images and words referring to meat break into the language of vegetarianism and the same rhetorical weapons usually adopted to say "eat meat" are now employed to sustain a vegetarian diet. For all of those interested in the production of meaning through images and words, this is certainly of great relevance. From the point of view of this book, and considering that this study also clarifies the reason why all of this happens, it is clear that this study may enlarge the boundaries we have assigned to the rhetoric of vegetarianism so far, adding new strategies and processes of signification.

More than 30 years ago, in an illuminating study Gvion-Rosenberg (1990) investigated the presence of the word "hamburger" in the menus of many vegetarian restaurants in New York. She concluded that the reason for this was mostly due to a sort of loyalty to structured forms of eating which are hard to eliminate. Meat was a consistent part of these forms and vegetarian menus had to submit to this.

DOI: 10.4324/9781003039013-10

Gvion-Rosenberg (1990) is certainly right when she sees meat as a sort of necessity in our diet and in the social and cultural discourse revolving around eating. The centrality of meat in our society has been underlined by many studies which have focused on the various connections which this food has toward many aspects of our lives.

Nick Fiddes (1991) argues that meat is a powerful, symbolic element regulating many interpersonal relationships. It confers masculinity, reminds people of blood, and contributes to constructing the supposed human supremacy over animals. Noëlie Vialles (1994) focuses on the process of slaughtering animals and the cultural separation that our society operates between the image of the living animal and the meat we eat. Carolyn Marvin and David Ingle (1999) link the experience of eating meat to the soldier's "touching death."

Carol Adams (2010) considers meat as a symbol of male violence to women and compares the violence over the animal to the violence against women. Joy (2010) points out that consumerist and neoliberal societies construct a system of beliefs based on power that encourages people to eat meat, meant as a cultural product. Finally, from a historical perspective, I have demonstrated that meat also shaped Nazi propaganda and the hatred against the Jews; and Fascist idealization of sacred vegetarianism (Buscemi 2016, 2018).

Moreover, meat has always had an important role in studies that have addressed it either as a function of something else or as an element structuring societies. In the first group, Norbert Elias (1978), Jack Goody (1982), and Stephen Mennell (1996) have found how the manners relating to eating meat have been part of a sort of "civilizing process" (Elias 1978), a virtuous and continuous improvement of the human behavior. The second branch of research is mostly represented by Claude Lévi-Strauss (1961), who finds that the way meat is cooked structures social relationships and discourses on power.

Interestingly for this study, discourses on vegetarianism and veganism have always been seen as in contrast with eating meat. Since Peter Singer's (2015) seminal study, the two practices have been considered as two fields speaking two different languages. Among the many, Gruen and Jones (2016) point out that rejecting meat is an ethical discipline aiming to avoid animal sufferance. Engel (2013) lists all the health advantages of a diet abstaining from meat. Gaard (2002) finds vegetarianism as a third-generation concept liberating women, after feminism and ecofeminism. Ali (2015) points out that vegetarianism is key to avoiding men subjugating women in Islamic countries.

Barboza (2019) links morality to eating or rejecting meat by analyzing marketing campaigns on food. Lonkila and Kaljonen (2022) find the marketing of meat alternatives in transition from distancing to approaching to meat language. Elzerman et al. (2011) have found that when it comes to choosing meat alternatives, the shape and appearance relating to meat in these products count more than flavor and taste. Lacy-Nichols et al. (2021)

have analyzed the health claims on vegetarian food packaging and have noticed that they often mention protein and other ingredients to position their products in relation to meat. Rödle (2018) points out that vegetarian food packages insist on the same myths that are usually referred to meat, stating that eating these foods is normal, natural, and necessary. Bogueva and Marinova (2019) focus on meat alternatives and masculinity and find that marketing should do more to reconcile these two poles, often considered as opposite.

Since Gvion-Rosenberg's (1990) and many of the studies above mentioned have been published, many things have occurred in the vegetarian scenario. Refusing to eat meat has become widespread in many parts of the world, while, on some parts of the planet, eating meat has been decreasing. Vegetarian foods have invaded the shelves of the most popular supermarkets; many celebrities are vegetarians and suggest rejecting meat but many veggie products continue to mention the enemy in their name, even though meat is clearly absent as an ingredient. In brief, meat's necessity is always there. The persistence of meat suggests adjourning Gvion-Rosenberg's (1990) approach and trying to understand why we are still there with vegetarian and vegan products containing in their names the terms "sausage," "chicken," and so on. It seems that two opposite poles such as eating meat and rejecting it speak the same language; or, at least, that the second has adopted the language of the first. These are the issues that this chapter revolves around.

Theoretical Framework: From Modernity to Postmodernity

The theoretical starting point in Gvion-Rosenberg's (1990) study was semiotics. She started from Peirce's (1960) classification of a sign as an icon, an index or a symbol, or, better, as embodying in turn one of the three categories. For Peirce (1960), a sign is composed of a signifier (its form, such as a word or a sound), and a signified (the concept or concepts that it represents). A signifier represents its signified in three different ways: it may be an icon, when it represents physically and directly its signified (a portrait); an index, when it recalls its signified (the smoke for fire); and a symbol, when there is no resemblance and the connection is culturally shared (the color red for love).

Gvion-Rosenberg (1990) found that in the analyzed New York menus the original meat hamburger was an icon, as it represented itself; and that the vegetarian hamburger was an index, as it somehow showed evidence of the concept of the hamburger. Starting from here, this research wants to adjourn the iconic, indexical, and symbolical relationships between the analyzed signifiers (vegetarian and vegan foods recalling meat) and their signifieds.

To do so, we have to understand the difference between the modern and the postmodern world. What Peirce theorized, and Gvion-Rosenberg applied, relates in fact to modernity, while in the postmodern scenario the processes of representation and signification occur differently.

Talking about propaganda, Bauman (1997) considers communication in the modern age as based on a neat distinction between purity and impurity. In the modern scenario, the two concepts are opposed to each other and neatly separated, belonging to two different spheres. Conversely, in the postmodern era, this separation disappears and purity and impurity tend to interrelate. What is key to this process is consumerism. Bauman (1997) advances that in postmodern times, purity means being successful in the market, winning the commercial competition whatever the strategy. In the end, those who remain excluded from consumption are impure.

Another element of postmodernism is the modification of the process of signification. Baudrillard (1994) advances that in this era reality, or at least what we intended before for it, tends to disappear and that the signifier may frequently change its signified. Baudrillard (1994) maintains that the new reality is what the media represent, which is different from what we once considered as "reality." He follows on from a short story by the Argentinian novelist Jorge Louis Borges, where an Emperor asks a group of cartographers for a very detailed map of the empire. The cartographers make maps more and more detailed but the Emperor always asks for major precision. In the end, the map is detailed, but it covers the territory completely, hiding the real conformation of the empire, and prefiguring a new reality which has cancelled the previous. People are thus convinced that the map is actually the real territory.

Baudrillard (1994) says that this is exactly what happens with postmodern media. They have replaced reality, and people believe that what the media represent is actually what exists. What the media create are simulacra of reality, mediated models of it. Reality, or at least the old one, has disappeared, and the media have created a new one. Paradoxically, these simulacra tell us something true, which belongs to our times. Baudrillard (1994) mentions Disneyland, a simulacrum which certainly represents a non-existing world, but that suggests much about the lives we live. Media representations, thus, contain true elements more than the old reality when we want to understand the present world.

The concept which helps us comprehend this shift in signification is re-mediation. Bolter and Grusin (1999) suggest that media representations are continually adjourned and re-invented by new forms of expression and new channels of communication. Each new medium or form of expression, in fact, takes something from the older one and re-elaborates its process of signification and its content in relation to the new social and cultural context. In this way, they construct new messages based on already existing styles and representations. What becomes old somehow survives by signifying other concepts.

It is a kind of mediation of a previous mediation, a continuous flux in which one meaning replaces the other. Importantly, Bolter and Grusin (1999) argue that this is a necessary process for the media, especially the postmodern ones. Even though re-mediation has affected communication in every age, the postmodern way of communicating is profoundly based on it. Re-mediation has its risks. The most important one for this study is the loss of contact between signifier and signified. As Bolter and Grusin (1999) say, the continuous multiplication of meanings weakens the adherence of the message to reality. Thus, on the one hand re-mediation guarantees the existence of communication but, on the other hand, it weakens its informative function or, like Baudrillard's simulacra, re-invents reality.

Finally, the alteration of the relation between signifier and signified sometimes ends up in adopting the language of the enemy. In Buscemi (2020) I have already analyzed this phenomenon in the context of propaganda, demonstrating that this practice has sometimes the aim of taking possession of the enemy, acquiring their credibility and finally replacing it. This is another relevant lens useful to guide the insight of the researcher in this study. In the end, the question which this study wants to answer is: through which communicative strategies do big retailers re-mediate meat to sell vegetarian and vegan foods recalling meat on their websites? The next section explains the methods adopted to answer this question.

Methods

First of all, it is necessary to explain that in this study, analyzing vegetarian products recalling meat means investigating textual and visual elements characterizing their packages and promotional web pages. Thus, it means that this study is based on the analysis of language, writings, packaging, design, colors, shapes, etc. All of this is seen here as rhetorical and communicative elements creating meanings, rhetorical strategies adopted to address the consumer.

Methodologically, the chapter applies social semiotic multimodality, which "aims to shed light on how people use modal resources in a certain social context" (Fortanet-Gómez and Crawford Camiciottoli 2015, 1). Semiotic analysis has long been seen as a method excessively detached by the concrete world where the analyzed phenomena occur; however, its social version has demonstrated impressive attention to the social and cultural contexts which surround the investigated fields. In this regard, Van Leeuwen (2005) maintains that social semiotic analysis constitutes a relevant form of insight both into "the material resources of communication and the way their uses are socially regulated" (Van Leeuwen 2005, 93).

Going further into the details of the method, for Van Leeuwen (2005), social semiotic analysis is made up of four objects of investigation: discourses, genre, style and modality. The term "discourses" does not refer only to speeches or written texts, but also to "the socially constructed knowledges of some aspects of reality" (Van Leeuwen 2005, 94). They are thus large topics shared by a community of people. For example, "war" may be a form of discourses, and discourses on the war in a specific context (such as a nation, a political party, etc.) may be analyzed in newspapers, films, videogames, etc. Discourses do not belong to just one perspective. War may be seen differently and from various points of view, and the analysis should take into account this variety.

The second element theorized by Val Leeuwen (2005) is genre, which refers to "templates for doing communicative things" (Van Leeuwen 2005, 128). Genre is thus constituted by sorts of containers that include specific forms of communication. The Western film is certainly a genre, like advice columns in magazines, or crime TV series.

The third element is style, and it relates to how signs are used to represent identities, ideologies, values and other concepts. For example, the way a package is designed, the layout of a magazine or the way a series is filmed may be considered as belonging to the element of style.

Finally, "modality is key to studying how people use semiotic resources to create the truth or reality values of their representations, to communicate, for instance, whether they [the messages] are to be taken as fact or fiction, proven truth or conjecture" (Van Leeuwen 2005, 91). It may thus involve the strategies adopted in order to underline the degree of truth of a message, its credibility, and its reliability. In other words, for social semioticians, a message cannot secure the absolute truth, but we can study how a message has been constructed to seem true (Hodge and Kress 1988; Kress and Van Leeuwen 1999; Van Leeuwen 2005; Piga 2018). Van Leeuwen (2005) also underlines that these four categories may be used either all together or in turn.

More specifically, in the case of this study, the four of Van Leeuwen's categories will be applied in the following way: discourses are the topics that emerge by the analysis of these products. Genres are the wider categories which these products belong to, and they may involve food categories (meaty foods, vegan products, etc.) or wider categories such as British food culture, environmentalism, etc. Style comprises expressive elements such as the layout of the analyzed websites, the colors present on the packages of these products, the fonts which are used, etc. Finally, modality corresponds to the resources adopted in the representations of these products to underline their degree of truth and credibility, develop the identification between the reader and what is sold and reinforce the credibility of the promotional messages of vegetarian and vegan products recalling meat.

Moreover, multimodality means that the items to be investigated are the result of the combination of different modes (drawings, photos, texts, colors, shapes, etc.). Thus, the researcher has to isolate the various modes constituting the analyzed item and look at each of them separately; finally, they have to gather them in a conclusive reading (Taylor 2014). This is what this chapter does, analyzing the many components of the packages of vegetarian foods recalling meat. Finally, in terms of sampling, this study has selected three countries that have really different approaches to food. Among them, this study has selected the US, the UK, and Italy as they have already been analyzed by the researcher as they are part of a wider project aiming at comparing these three forms of food culture. This study has chosen the biggest retailer in each country, as the aim is to investigate the promotion of these products not in niche markets but on mass retailers.

As regards the selection of the websites' pages, it is important to say that this study is based on qualitative research. This means that it does not aim to provide exhaustive generalizability (Newell and Burnard 2011; Ravitch and Carl 2016). Rather, it tries to identify, investigate and reflect on a specific group of packages of vegetarian and vegan foods recalling meat, aiming to achieve "'depth' rather than 'breadth'" (Blaxter et al. 2006: 64). Relating to this, the research has sampled a limited number of items according to purposive sampling, which is a strategy in which "sample units are chosen "purposively" for the ability to provide detailed understanding" (Ritchie et al. 2003, 107). This kind of sampling suggests choosing items based on a specific "characteristic that is known or expected to be salient to the research study" (Ritchie et al. 2003, 107). In this case, the characteristic is the presence of rhetorical and communicative strategies based on remediation of meat.

To carry out the study, the researcher has thus searched the websites of the three biggest retailers, respectively, in the US, the UK, and Italy, which are Walmart, Tesco, and Conad, respectively; and has chosen the vegetarian foods recalling meat which presented in their communicative strategies forms of remediation of meat. This study aims to find out how the products sold by the most popular and powerful retailers in three different contexts adopt rhetorical strategies by subverting the language of meat in the market segment of vegetarian food. The products that have been analyzed are: the US MorningStar Farms Plant-Based Vegan Meatless Original Meatballs, with soy and wheat; Tofurky Plant-Based Italian Original Sausage, containing tofu, wheat and soy and recommended with spaghetti; and MorningStar Farms Plant-Based Vegan Meatless Original Chicken Nuggets, based on soy and wheat and corn flour. The British Vivera Plant Meatballs, made up of soy and wheat; Richmond meat-free sausages, also containing chicory and coconut oil; and Vivera Veggie Greek Kebab, which reinvents kebab thanks to soy and rapeseed and sunflower oil. Finally, the Italian Burger di soia classico biologico, which replaces meat with soy; and Nuggets alle verdure biologici which adopts the term nuggets which usually refers to chicken.

These products have been isolated and analyzed by investigating the single modes which they showed. The analysis has connected these modes to the four Van Leeuwen's categories. At the end of this stage, more general results regarding all of the modes and the entire research have been made explicit.

The Analysis

Discourse

The most frequent discourse stressed on the Walmart website is taste. Both the descriptions of the products and the comments of the consumers very often refer to the flavor of these foods. Walmart website reports the description of the Veggie Meatballs by MorningStar Farms, which suggests: "These tasty meatballs are perfect for any night of the week or special occasions with friends and family" (Walmart 2020a). Similarly, on the same website, Tofurky Original Sausage Italian is considered perfect to prepare "a hearty and delicious meal," as it "stands out with flavors of fragrant basil and sweet sun-dried tomatoes (Walmart 2020b). Moreover, it is stated that Veggie Meatballs contribute to "preparing an exciting meal for a crowd, MorningStar Farms Veggie Meatballs are sure to delight vegans, vegetarians, and meat-lovers alike" (Walmart 2020a), enlarging the potential target of the product.

Thus, the taste is the key through which meatless products recalling meat are re-mediated. The promoted taste is the same as meat, and this is thus the way that these vegetarian products are represented to substitute foods with animal origins. As in any case of remediation, an old representation, in this case meat, is represented in a new way. Meat is here the simulacrum of meat, as Baudrillard would say.

Interestingly, on the Tesco website, the discourse on taste is much less present, while other issues are the focus of the promotion. In Tesco (2020a), immediately after the list of the ingredients, there is the writing "Change the world." Taste is a secondary element when it comes to suggesting this product, which is strongly recommended for more social and environmental reasons, as the following text hints: "We believe life is better when you eat less meat. So, we want to help people make their change with our delicious plant-based meat. Join us and feed the Goodness Revolution!" (Tesco 2020a). The promotion of the Richmond Meat Free Sausages shifts the focus toward tradition: "It's fair to say we know a thing or two about sausages … and we should do … we've been proudly making them for over 100 years! We have now created our very own meat-free sausages, staying true to our Irish recipe everybody loves" (Tesco 2020b).

However, on the Tesco website, what is lost in comparison with Walmart is rescued in the comments of the consumers. A Tesco customer underlines in fact: "This tasted so good! Having tried Vivera mince, I was not disappointed with the meatballs. Really tasty!" (Tesco 2020a); and among the more than 250 reviews of the Richmond sausage, many proudly refer to the taste of the product. "Absolutely as yummy as you can get but healthy – for a sausage!" (Tesco 2020b). Another customer adds: "They look smell and taste just like the meat version, minus the dodgy bits you sometimes get in real meat" (Tesco 2020b), and by saying so removes the focus on tasty and real meat replacement.

On the US website, it is clear that the remediation of meat through the promotion of meatless products recalling meat serves the purpose of replacing the pleasure of meat; conversely, in the UK it aims to improve the environment and change the world. Meat is re-mediated to promote and sustain a political and social stance. This will be examined in more depth in the conclusive section, but this study needs to highlight this aspect already at this point, as it is a fundamental difference between the two communicative approaches.

In the Italian case, vegetarian and vegan products recalling meat are inserted in a wider food series called Verso Natura (toward nature) (Conad 2020a). The series is for vegetarians, vegans, or simply people who desire to eat fewer animal proteins without giving up flavor and taste. It is defined as containing products that address people who want to love themselves and the world. The veggie series is in fact a choice of wellness, which people can live with freedom and flexibility.

As regards Nuggets alle Verdure Biologici (organic vegetable nuggets), it is stated that they are made with organic ingredients, do not contain GMOs, and respect the environment in each stage of the production process (Conad 2020b). Moreover, the promotional text says that their main ingredient is tofu, which is a millenary food of Asian origins, containing many proteins and a little salt. Later, the origin and the production process of this item of food are explained.

Similarly, in Burger di Soia Classico Biologico (organic classic soy burger) (Conad 2020c), a great part of the promotional text is devoted to the historical explanation of what soy is. It is in fact underlined that soy is a pulse of Asian origin that has been adopted in the kitchen since antiquity. It is considered one of the vegetables which are richest in proteins and which contains minerals and essential amino acids.

While the US and the UK retailers' websites overtly constitute a case of remediation of meat, although, as already said above, in two different directions, in the Italian case this actually does not happen. Vegetarian products recalling meat, in fact, here are products per se, not showing any connection to meat apart from their names, which, however, are more detached from foods of animal origins than the American and the British ones. While in fact in the first two cases the names of the products directly refer to animals (chicken, beef, etc.), in the Conad website they only refer to meat dishes (burger and wurstel). In the end, the link to meat is much weaker here, and the products seem to be an independent and new category of foods. This anomaly is confirmed when we look at the second of Van Leeuwen's categories, that is, genre.

Genre

What strikes the viewer of the Walmart website is the continuous reference to the real meat, intended here as a broader food category. In the case of the vegetarian meatballs (Walmart 2020a), for example, this is clear in the way the product is presented on its package. First, the images of the meatballs resemble meat in a really impressive way. The photos are just taken to make the product similar to the "real" meatballs, and thus connecting the meatless to the meaty one. Second, on the package, the term "veggie meatballs" is written with very different fonts. The word "veggie" is really small, while the word "meatballs" is big and decidedly more evident. This strategy aims to underline the vicinity of the product to meat and meaty products, leaving a secondary role to its vegetarian and vegan character.

This also happens on the page dedicated to Veggie Chick'n Nuggets (Walmart 2020c). Again, the images of the products overtly resemble meat and on the package, the word "veggie" is really small and the term "Chick'n Nuggets" is conversely big and highlighted. Moreover, here the text underlines that this product will also "please meat-lovers," further bridging the gap between products that do not contain meat and products that do.

In conclusion, it is interesting to note that, even though the products are meatless, there is a continuous relation to the food of animal origins. In the end, the promoted vegetarian products are represented as belonging to the wider genre of meat, as a subgenre of products containing meat, even though they are meatless.

This does not happen on the Tesco website, where meaty products are not recalled either visually or textually. The Vivera Plant Meatballs (Tesco 2020a) are not photographed as similar to meat and the text, as said above, focuses on other issues. Rather than on meat, the writing centers on the revolutionary character of these items of food and invites the consumers to "change the world byte by byte" (Tesco 2020a). In the Richmond meat Free Sausages (Tesco 2020b) this social aim is reinforced by suggesting customers follow the company on Facebook and Instagram. The invitation wants to contribute to building a community and make the consumers feel a part of it.

The social and political character of this kind of food is also on the page promoting Vivera Veggie Greek Kebab (Tesco 2020c), where it is stated that "Kebabs can save us," as it is "100% plant-based." While in the Walmart website the genre of these products is meat, here the category which the analyzed foods belong to is something concerning the ability of food to change the world for the better. We can define it as the change-the-world food genre, which in Britain had a very popular representative in the celebrity chef Jamie Oliver, who in many programs strove to demonstrate that healthy foods may improve the lives of schoolchildren, disadvantaged youngsters, the poor, and people in general.

Even more different is the case of Italy. In its stressing the organic character and the historical origins and usage of soy and tofu, the Conad website puts the connections to meat and social aims as a secondary role. Rather, the Italian retailer centers on the specific characters of the product. However, even in this case, there is the reference to a wider category, helping construct a more precise identity of what is sold. The promoted Burger di Soia Classico Biologico (Conad 2020c), is suggested through promoting its being organic, the absence of GMOs among its ingredients, and its being rooted in history and tradition. Finally, it is promised that by eating it people would improve their wellness first, and the environment, second.

The same thing occurs on the page about Nuggets alle Verdure Biologici (Conad 2020b). The references are the same as in the previous case: the promoted product is to be consumed as it is organic, does not contain GMOs, and is rooted in history and tradition. These are thus the elements that the Italian promotion is based on.

All of these characteristics are the same principles supported by the Slow Food movement. Slow Food is a powerful agent in the field of Italian food culture, insisting on consuming organic foods, without GMOs, and deeply rooted in history and tradition. Certainly it does not represent a break in comparison to Italian food culture in general; rather, the movement is the most popular agent prompting principles and rules which have always been present in the Italian approach to food, and has been able to renovate it without subverting older Italian food habits. The genre which the product promoted on the Conad website refers to, thus, is Italian food culture. Instead of referring to real meat or social change, the Italian product's promotion is based on following the basic principles of the dominant way that the Italians look at food.

Style

In terms of style, it is interesting to note that the Walmart website adopts words that emotionally address the reader underlining the goodness of the product. In the page of the Chick'n Nuggets (Walmart 2020c), for example, just in the first line of the text there are adjectives such as "delicious," "lightly seasoned," "crispy," and "crunchy." In the case of the vegetarian Italian sausage (Walmart 2020b), the same language is adopted. In fact, it mentions the "flavors of fragrant basil" and "sweet sun-dried tomatoes." However, a hint at Italy is added. The text says in fact that "they bring a smile to spaghetti" and, in line with the above mentioned Italian food culture and Slow food principles, it is underlined that the product is GMOs free.

Visually, while the different dimensions between the words "veggie," really small, and "Chick'n Nuggets," very big, have already been explained, the packaging of the products presents interesting details. In the case of the MorningStar Farms series (Walmart 2020a, 2020c), it is composed of two different nuances of green and a big photo of the product with large writing. In the case of Tofurky Sausage (Walmart 2020b), produced by another company, it is evident the dimension of the term "sausage," which is the biggest one, and is followed in dimension by the words "original" and "Italian," while the terms referring to the absence of meat and vegetarian and vegan ingredients play a secondary role, being much smaller.

In Britain, the Tesco website presents the Richmond Meat Free Sausages (Tesco 2020b) declaring "we love to chat, drop us a line" and writing the postal address of the company. In general, in the website, the cooking instructions are much more developed than the US and the Italian analyzed websites, and this is another way of keeping in contact with the customer, like the already mentioned invitation to follow the company on its social media accounts. Even in this case, the dominant color is the green one, but the writing is in an old-style font and suggests that the product comes from an original Irish recipe. Similarly, the plate and the table in the photo resemble plates and tables of the countryside houses of the past.

The packaging of the Vivera series (Tesco 2020a, 2020c) is much more similar to the US packaging of the MorningStar Farms products. The green is by far, the principal color, the writing is big and clear but there is less difference between the dimensions of the terms "plant" and "meatballs" (Tesco 2020a). In the case of the Plant Greek Kebab (Tesco 2020c), it is "Greek" which is the biggest word, while "plant" and "kebab" are almost of the same size. In this series, finally, there is no hint at old-style writings and images which constitute the background of the promoted product.

The style connoting the products sold on the website of the Italian retailer Conad presents interesting differences. We have already said that the principles which are stressed are the organic nature of the food, the absence of GMOs, and the root in history and tradition. The adjectives relating to flavors and taste (crispy, crunchy, etc.) present on the American website are absent here. The description of the product is sober and full of pieces of information about the ingredients and their origins. It is underlined, for example, that soy is 100% Italian (Conad 2020b).

Visually, the packaging of these products is different from those analyzed on the American and British websites. The green color is almost absent, while the white is dominant. The fonts are thin and never too big and flashy. The lettering is small and clear, and there are bright colors and large empty spaces. Even the rest of the layout is composed of really thin lines. All of this, along with the white background and the large empty spaces, totally absent in the other analyzed cases, transmits a sense of lightness, cleanliness, and purity, less informative (the texts on the packaging are really short) but more emotionally involving.

Modality

One of the strategies that the Walmart website adopts to reinforce the credibility of the product is to leave considerable room for the cooking instructions, as in the case of Walmart (2020b), which is a frozen product and thus would not need such detailed suggestions. However, explaining how what is in the package may become a flavorful dish certainly reinforces the credibility and the actuality of the product. Another strategy to reinforce the sense of concreteness of the sold food is to suggest physical places, that is, stores, where the product can be bought. In this way, the sold item of food is certainly perceived by the customers as something existing and concrete.

However, these are strategies that are adopted by Walmart for each kind of product. To better understand how the communication of the product reinforces the identification between the customer and what is sold, it is really useful to read the comments of the public. The website highlights one of the reviews, which is considered the "most helpful" (Walmart 2020c), while the others follow in the chronological sequence below the highlighted one. Meaningfully, in Walmart (2020c) the most helpful review is titled "Like Real Chicken!" and in the text the customer explains all the reasons why the veggie chicken is like the real one, to conclude that this product is even "better than Tyson chicken nugs" (Walmart 2020c). Another review is "100% like Chicken," and the reviewer says: "Serious you would never know they were not chicken. They taste 100% like a tyson chicken nugget" (Walmart 2020c).

We have already seen that positively comparing veggie products recalling meat to real meat is a general strategy of the Walmart website. However, leaving this role to a reviewer (regardless of the credibility of the review, whose verification is not part of this study), and thus allegedly to ordinary people who are not directly connected to the company, adds more credibility to the product.

Even on the Tesco website, cooking instructions bridge the gap between the product and the customer, making the photographed foods something more concrete. However, as in the US case, what is interesting to analyze in terms of Van Leeuwen's modality are the comments of the (presumed) customers. About the veggie kebab (Tesco 2020c), for example, it is interesting that the reviews are written in turn by vegans or vegetarians and by people who are meat-eaters only wanting to reduce their meat consumption. Another reviewer suggests this product to people who are becoming vegetarians. Clearly, all of this favors the involvement of people with different approaches to meat, developing more types of identification between customers and the product.

In another review, the credibility of the product as an almost form of meat is reinforced by saying: "Everything you love about kebab minus all the bits you don't (the grease)" (Tesco 2020c); and later: "The whole family have eaten them really impressed (Tesco 2020c). Again, these texts may have a great value in involving the reader and improving their perception of the veggie kebab, which is now more concrete and true than before, when it was a product only visible externally in its packaging and supported by a promotional text.

Many other reviews on the Tesco website, regarding this and the other analyzed products, follow this trend. For example, veggie sausages are reviewed as "absolutely delicious, taste exactly the same as the ones with animal fats in but plant-based, much better for our planet!" (Tesco 2020b). The review is titled "Absolutely no compromise on flavour" and interestingly on the one hand it bridges the gap between meatless and meaty products, and on the other hand it stresses the social and political hint that has already been seen on the British website.

All of this disappears on the website of the Italian retailer Conad, which does not admit reviews or other forms of interaction with the customers. Similarly, as said, the website never indulges in comparing the taste of meat to that of meatless products to give the reader a sense of credibility. All of the communication of these pages is based on what has already been analyzed, that is, the feelings of belonging to the glorious tradition of Italian food culture, and obeying to its principles, which are the same stressed by the Slow Food movement.

The interesting novelty in this traditional scenario is a more open approach to food coming from other countries. Italian food culture has always been viewed with suspicion cuisine coming from other countries. In this case, instead, the Asian origins of tofu and soy are overtly stated and taken as an example of high quality. These "foreign" ingredients are somehow legitimized and considered as foods certainly coming from abroad but positively adopted by the Italian tradition: an open approach, which is mediated by the statement that these products, although of foreign origins, are produced in Italy. They are thus partly Italian. Apart from this nuance, this open approach is quite common in the Anglo Saxon countries but that has rarely been adopted in Italy, whose approach to food has often been considered nationalistic. In the end, thus, the sense of truth and credibility of these products is entirely based on the fact that they pertain to the category of Italian food.

Conclusion

The analysis which has been carried out so far demonstrates that there are three different ways of rhetorically representing items of vegetarian and vegan foods recalling meat. The three analyzed websites, in fact, represent these products in three different ways in relation to the real meat and rely on different strategies to draw the consumer's attention and appetite. These different perspectives, as we will see below, are also reflected in the different semiotic categories of icon, index, and symbol (Peirce 1960).

On the Walmart website, real meat is an always present entity. Meat is continuously compared with the analyzed products. As a result, real meat becomes a close relative of these products, which would never exist without foods of animal origins. The meatless products, thus, in their recalling meat very often pay a kind of tribute to the real meat, which is always there and which the meatless products may only look like, in an attempt to imitate the original, as they appear in the photos. Semiotically speaking, and returning to Peirce's (1960) icon, index, and symbol, meatless products are a sort of indices of real meat. Like smoke for fire, they remind us of the existence of a meaty sausage or nugget. As in Gvion-Rosenberg (1990), real meat is still an icon, as it represents itself; while vegetarian products recalling meat are indices, as they show evidence of the concept of meat.

Tesco's website behaves differently. Here real meat is not obsessively present as on the American website. Conversely, meatless products are represented in contrast with real meat. There is much less comparison between meatless and meaty food. Only the comments of the customers hint at it, while the texts promoting the vegetarian and vegan products do not consider their meatless products as an imitation of real meat. Rather, they focus on something else, that is, a sort of opportunity to change the world. Tesco's communication is based on the fact that it does not matter about the flavor of these products. What is important is that they can replace the "old" meat, producing a change in the world and an improvement in the conditions of the planet. In this sense, these products are no more indices of meat, as for Walmart's website. The signification shifts from the close link to real meat toward a culturally shared connection to environmental ideals. These products become thus a symbol of social change.

Finally, on the Italian Conad website, real meat is absent, apart from the names of the products which recall meaty dishes. The rhetorical strategy adopted by Conad is neither "it is like meat," as for Walmart, nor "it can change the world," as for Tesco. Rather, the communication of the product is based on the belonging to Italy and its food traditions. The products do not have other products to be compared with or social and political ideals they are aspired to. They only represent themselves and are thus icons, signifiers physically and directly representing their signifieds.

What has been said so far also impacts Bauman's (1997) view of the postmodern division between purity and impurity. In the traditional vegetarian discourse, abstaining from eating meat means being pure, in contrast with the impurity of eating animal foods. As in Bauman (1997), the division between the two concepts here is much weaker. By choosing words regarding meat (sausage, chicken, kebab, etc.), these veggie products somehow speak the language of the enemy (Buscemi 2020). By doing so, the differences between the two opposite poles meaty/meatless fade and purity and impurity confuse with each other. As foreseen by Bauman (1997), the real battlefield in postmodern times becomes the market; and in fact, it is not by chance that these representations occur on the websites of three mega retailers. Using the language of the enemy here has aimed to commercially expand the influence of a new trend, abstaining from meat, and selling it by stealing from a traditional food habit such as eating meat. Sausage and even chicken or beef are now not an exclusivity of the carnivorous world. Now they also pertain to the vegetarian and vegan vocabulary. All of this, it is important to repeat, happens on the battlefield of the market.

By selling veggie products recalling meat, these companies have constructed three different simulacra which, as in Baudrillard's (1994) map, hide reality. But as in Baudrillard, the reality represented by the simulacra ends up being truer than the original, as it says something about the new world. As with Disneyland, which represents a very true non-existing world, these veggie products named "chicken nuggets" or "sausage" represent non-existing meaty dishes. It is through Van Leeuwen's (2005) four categories, especially modalities, that we have seen how the three websites have constructed their new versions of reality through a process of remediation of meat.

Saving the names of meaty dishes has allowed them to change the meaning, in a process through which each signifier has changed its signifier. In Walmart, meatless products have become another form of meat; in Britain a means through which we can change the world; and in Italy a further example of the purity of Italian food culture. As in any form of remediation (Bolter and Grusin 1999), a new form of representation has taken something from an older one and has re-elaborated its signification. Based on already existing representations (meat), they have constructed new messages (the promotion of meatless products recalling meat) according to new cultural and social contexts (popularization of vegetarianism and veganism).

The linguistic subversion demonstrated by the semiotic analysis has its reasons, and as said at the beginning, finding them is one of the aims of this chapter. The three retailers and the companies producing these foods have re-invented language to reach a portion of customers who want vegetarian products without forgetting meat: in short, to sell more food and earn more money, in line with what has been said by Bauman (1997). Thus, ethical considerations and environmental aims here have served the purpose of increasing income. Certainly, this is not the first case where this happens, but what is relevant is the strategy to do it, that is, subverting the meaning of signifiers to sell a new and ethical signified. As a result, and as theorized by Bolter and Grusin (1999), signifier and signified have lost contact and we have on the shelves of our supermarkets chicken without chicken and sausage without sausage. The message does not adhere to reality anymore and the reason is the first aim of neoliberalism: commerce. Reality, as prefigured by Baudrillard (1994), has been re-invented to signify its contrary, to the point of creating a vegetarian chicken.

Bibliography

Adams, Carol J. 2010. *The Sexual Politics of Meat (20th Anniversary Edition): A Feminist-Vegetarian Critical Theory.* New York, NY: Continuum International Publishing.

Ali, Kecia. 2015. "Muslim and Meat Eating: Vegetarianism, Gender, and Identity." *Journal of Religious Ethics*, 43 (2): 268–288.

Barboza, Stephanie Ingrid S. 2019. "Equilibrium of the Food Marketing System: A Debate of an Ethical Consumption Performance Based on Alternative Hedonism." *Food Ethics*, 2: 139–153.

Baudrillard, Jean. 1994. *Simulacra and Simulation.* Ann Arbor, MI: The University of Michigan Press.

Bauman, Zygmunt. 1997. *Postmodernity and its Discontents.* Maiden, MA: Polity.

Blaxter, Loraine, Hughes, Christina, and Tight, Malcolm. 2006. *How to Research*, 3rd ed. Maidenhead, UK: Open University Press.

Bogueva, Diana and Marinova, Dora. 2019. "Reconciling Not Eating Meat and Masculinity in the Marketing Discourse for New Food Alternatives." In *Environmental, Health, and Business Opportunities in the New Meat Alternatives Market*, edited by Diana Bogueva, Dora Marinova, Talia Raphaely and Kurt Schmidinger, 260–282. Hershey, PA: IGI Global.

Bolter, J. David and Grusin, Richard. 1999. *Remediation: Understanding New Media*. Cambridge, MA: MIT Press.

Buscemi, Francesco. 2016. "Edible Lies: How Nazi Propaganda Represented Meat to Demonise the Jews." *Media, War and Conflict*, 9 (2): 180–197.

Buscemi, Francesco. 2018. "The Sin of Eating Meat: Fascism, Nazism and the Construction of Sacred Vegetarianism." In *Proteins, Pathologies and Politics: Dietary Innovation and Disease from the Nineteenth Century*, edited by David Gentilcore and Matthew Smith, 137–147. London, UK: Bloomsbury.

Buscemi, Francesco. 2020. "Il Simulacro della BBC: Il Linguaggio della Propaganda Isis nei Filmati con John Cantlie." *Cinema e Storia*, 1, 145–162.

Conad. 2020a. "Verso Natura VEG." *Conad*, https://www.conad.it/nostri-prodotti/scelte-di-benessere/verso-natura-conad/verso-natura-veg.html?page=3.

Conad. 2020b. "Nuggets alle verdure Biologici 160 g." *Conad*, https://www.conad.it/nostri-prodotti/scelte-di-benessere/verso-natura-conad/verso-natura-veg/prodotto.8003170060975.html.

Conad. 2020c. "Burger di Soia Classico Biologico 2 x 90 g." *Conad*, https://www.conad.it/prodotti/dettaglio-prodotto.8003160036089.html.

Elias, Norbert. 1978. *The Civilising Process: History of Manners*, Vol. 1. Oxford, UK: Basil Blackwell.

Elzerman, Johanna E., Hoek, Annet C., van Boekel Martinus A.J.S. and Luning Pieternel A. (2011) "Consumer Acceptance and Appropriateness of Meat Substitutes in a Meal Context." *Food Quality and Preference*, 22 (3): 233–240.

Engel, Mylan. 2013. "The Immorality of Eating Meat." In *The Moral Life: An Introductory Reader in Ethics and Literature*, edited by Louis Pojman, 856–889. New York, NY: Oxford University Press.

Fiddes, Nick. 1991. *Meat: A Natural Symbol*. London, UK: Routledge.

Fortanet-Gómez, Immaculada and Camiciottoli, Belinda Crawford. 2015. "Introduction." In *Multimodal Analysis in Academic Settings: From Research to Teaching*, edited by Belinda Crawford Camiciottoli and Immaculada Fortanet-Gómez, 1–14. New York, NY: Routledge.

Gaard, Greta. 2002. "Vegetarian Ecofeminism: A Review Essay." *Frontiers, a Journal of Women Studies*, 23 (3): 117–146.

Goody, Jack. 1982. *Cooking, Cuisine and Class: A Study in Comparative Sociology*. New York, NY: Pantheon Books.

Gruen, Lori and Jones, Robert. 2016. "Veganism as an Aspiration." In *The Moral Complexities of Eating Meat*, edited by Ben Bramble and Bob Fisher, 153–171. Oxford, UK: Oxford University Press.

Gvion-Rosenberg. Liora. 1990. "Why do Vegetarian Restaurants Serve Hamburger? Toward an Understanding of a Cuisine." *Semiotica*, 80 (1–2): 61–79.

Hackett, Jolinda. 2019. "The Best Vegetarian and Vegan Meat Substitutes to Try." *The Spruce Eats*, August 19, https://www.thespruceeats.com/best-vegetarian-meats-and-substitutes-3377742.

Hodge Robert, Kress Gunther. 1988. *Social Semiotics*. Cambridge, UK: Polity Press.

Joy, Melanie. 2010. *Why We Love Dogs, Eat Pigs and Wear Cows: An Introduction to Carnism, the Belief System that Enable Us to Eat Some Animals and not Others*. San Francisco, CA: Conari Press.

Kress Gunther, Van Leeuwen Theo. 1999. *Reading Images. The Grammar of Visual Design*. New York, NY: Routledge.

Lacy-Nichols, Jennifer, Hattersley, Libby, and Scrinis, Gyorgy. 2021. "Nutritional Marketing of Plant-Based Meat-Analogue Products: An Exploratory Study of Front-of-Pack and Website Claims in the USA," *Public Health Nutrition*, 24 (14): 4430–4441.

Lévi-Strauss, Claude. 1961. "The Culinary Triangle," *Partisan Review*, 33 (1): 586–595.

Lonkila, Annika and Kaljonen Minna. 2022. "Ontological Struggle over New Product Category: Transition Potential of Meat Alternatives," *Environmental Innovation and Societal Transitions*, 42: 1–11.

Marvin, Carolyn and Ingle, David W. 1999. *Blood Sacrifice and the Nation: Totem Rituals and the American Flag.* Cambridge, UK: Cambridge University Press.

Mennell, Stephen. 1996. *All Manners of Food: Eating and Taste in England and France from the Middle Ages to the Present.* Champaign, IL: University of Illinois Press.

Newell, Robert and Burnard, Philip. 2011. *Research for Evidence-Based Practice in Healthcare.* Oxford, UK: Wiley-Blackwell.

Peirce, Charles S. 1960. *Collected Papers. Vol. II: Elements of Logic.* Cambridge, MA: Harvard University Press.

Piga, Antonio. 2018. "Epistemic Modality in the Visual Art of Dissent: The Expression of Realism in Ideological and Political Sardinian Murales." *Lingue e Linguaggi*, 28 (1): 227–251.

Ravitch, Sharon M. and Carl, Nicole M. 2016. *Qualitative Research: Bridging the Conceptual, Theoretical and Methodological.* New York, NY: Sage.

Ritchie, Jane, Lewis, Jane, and Elam, Gillian. 2003. "Designing and Selecting Samples." In *Qualitative Research Practice: A Guide for Social Sciences Students and Researchers*, edited by Jane Ritchie and Jane Lewis, 77–108. London, UK: Sage.

Rödle, Malte B. 2018. "Marketing Meat Alternatives: Meat Myths and Their Replication in Advertising for Plant-Based Meat." In *Handbook of Research on Social Marketing and Its Influence on Animal Origin Food Product Consumption*, edited by Diane Bogueva, Dora Marinova & Talia Raphaely, 327–343. Hershey, PA: IGI Global.

Singer, Peter. 2015. *Animal Liberation.* London, UK: The Bodley Head.

Taylor, Roberta. 2014. "Multimodal Analysis of the Textual Function in Children's Face-to-Face Classroom Interaction." In *Multimodal Epistemologies: Towards an Integrated Framework*, edited by Arianna Maiorani and Christine Christie, 228–244. New York, NY: Routledge.

Tesco. 2020a. "Vivera Plant Meetballs 200G." *Tesco*, https://www.tesco.com/groceries/en-GB/products/305978977.

Tesco. 2020b. "Richmond 8 Meat Free Sausages 336G." *Tesco*, https://www.tesco.com/groceries/en-GB/products/304381808.

Tesco. 2020c. "Vivera Veggie Greek Kebab 175G." *Tesco*, https://www.tesco.com/groceries/en-GB/products/303543032.

Van Leeuwen, Theo. 2005. *Introducing Social Semiotics.* London, UK: Routledge.

Vialles, Noëlie. 1994. *Animal to Edible.* Cambridge, UK: Cambridge University Press.

Walmart. 2020a. "MorningStar Farms, Veggie Meatballs, Original, 10.3 Oz." *Walmart*, https://www.walmart.com/ip/MorningStar-Farms-Veggie-Meatballs-Original-10-3-Oz/506903899.

Walmart. 2020b. "Turtle Island Tofurky Italian Sausage, 14 oz." *Walmart*, https://www.walmart.com/grocery/ip/Turtle-Island-Tofurky-Italian-Sausage-14-oz/173896638?athcpid=IID_173896638&athpgid=itempage&athcgid=null&athznid=null&athieid=null&athstid=CS014&athguid=466001f5-46cfa622-9c8f-80d69a18a716&athancid=null&athena=true&selected=true.

Walmart. 2020c. "MorningStar Farms, Veggie Chick'n Nuggets, Original, 10.5 Oz." *Walmart*, https://www.walmart.com/ip/MorningStar-Farms-Veggie-Chik-n-Nuggets-Original-10-5-Oz/49358456.

Index

For Product Safety Concerns and Information please contact our EU representative GPSR@taylorandfrancis.com
Taylor & Francis Verlag GmbH, Kaufingerstraße 24, 80331 München, Germany

www.ingramcontent.com/pod-product-compliance
Ingram Content Group UK Ltd.
Pitfield, Milton Keynes, MK11 3LW, UK
UKHW020139200926
475702UK00005B/819
* 9 7 8 0 3 6 7 4 8 2 7 9 4 *